D0987270

African Americans at Mars Bluff, South Carolina

African Americans

at Mars Bluff, South Carolina

Amelia Wallace Vernon

Louisiana State University Press
Baton Rouge and London

Manufactured in the United States of America
First printing
02 01 00 99 98 97 96 95 94 93 5 4 3 2 1

Designer: Glynnis Phoebe
Typesetter: G & S Typesetters, Inc.
Typeface: Bembo
Printer and binder: Thomson-Shore, Inc.

Library of Congress Cataloging-in-Publication Data
Vernon, Amelia Wallace, 1926–
African Americans at Mars Bluff, South Carolina / Amelia Wallace Vernon.
p. cm.
Includes bibliographical references and index.
ISBN 0-8071-1846-X (hard : alk. paper)
1. Afro-Americans—South Carolina—Mars Bluff—History. 2. Mars Bluff (S.C.)—History. 3. Rice—South Carolina—Mars Bluff. 4. Rice farmers—South Carolina—Mars Bluff. I. Title.
F279.M326V47 1993
975.7'84—dc20 93-15834
CIP

The paper in this book meets the guidelines for permanence and durability of the Committee on Production Guidelines for Book Longevity of the Council on Library Resources. ∞

To
Alex Gregg
(*1845–1938*),
Archie Waiters
(*1914–1990*),
and
all African Americans,
past, present, and future

Contents

Part Three: *Tom Brown*

Appendixes

Illustrations

Figures

Maps

Preface and Acknowledgments

I HAVE TRIED TO write this book simply and clearly so that others can enjoy meeting the African Americans at Mars Bluff and learning what they know. All scholarly material is reserved for endnotes and appendixes.

Many people contributed to this work. My indebtedness is not just to those named in the interviews but to all the people who pointed the way. They enabled me to see a world where people retained and used their African cultural heritage well into the twentieth century, a world too seldom recorded in print.

I am grateful that Archie Waiters lived and that he shared his story with me. Without him, this book could never have been written. Also, I want to express my gratitude to Peter Wood, Paul Richards, and Daniel C. Littlefield, who read my earlier attempt to tell the Mars Bluff story and gave criticisms that supplied a new sense of direction, the basis for this work. I have been fortunate to have had the skillful and insightful editing of Catherine Landry, managing editor, Louisiana State University Press, and the home-front support of Joyce Oberhausen and Jane Vernon. Deserving of special acknowledgment are Dinah Bervin Kerksieck for providing maps and Sidney Glass for selected printing of photographs.

To all of the people who have helped me, I am glad to have this opportunity to say how much I appreciate their kindnesses. I especially recall the courtesies shown to me by Tom Auge, Alpha Bah, Willie H. Bailey, Ruth Lee Bailey, Joseph Broadwell, Hazel Carter, Marion C. Chandler, T. T. Chang, Annie Coker, Leon H. Coker, Mary Scott Cooper, Lucas Dargan, Timothy G. Dargan, Dorothy Hicks Davis,

Lena Mae Douglas, Maggie Waiters Egleton, Sherry K. Ewert, Sidney Glass, Mattie Smalls Gregg, James Harwell, Lacy Rankin Harwell, Karen Hess, Roger Hux, Naida Humke, Frances Johnson, Nelson E. Jodon, Thomas L. Johnson, Charles Joyner, G. Wayne King, Melissa Leach, Iliya Likita, Daniel C. Littlefield, Barney Mattenson, Howard McNair, Elaine Nowick, Hazel O'Neal, Joseph A. Opala, Rubin Peterson, Atleene Pinkney, Paul Richards, Annie Lee Waiters Robinson, Cora Robinson, Dale Rosengarten, Horace Fraser Rudisill, Joseph W. Sallenger, Walter Sallenger, Rev. Frank Saunders, Richard R. Schulze, Mable Smalls Sellers, Joko Sengova, Anna S. Sherman, Suzanne Singleton, Isabell Daniels Smith, Kristen L. Smith, Lawrence Swails, John Martin Taylor, Harmon Taylor, Gordon Vernon, Tom Vernon, Catherine Waiters, Hester Waiters, Otis C. Waiters, Richard Waiters, Amelia M. Wallace, Annie P. Wallace, Mary Daniels Washington, Claudia Williamson Williams, Dorothy Smalls Williams, Ann Niemeyer Williamson, Juanita Cody Williamson, Matthew Williamson, Peter Wood, and Ida Ellison Zanders.

I am happy to thank the following publishers for permission to use excerpts from the works listed:

Greenwood Publishing Company, for excerpts from *The American Slave: A Composite Autobiography,* ed. George P. Rawick. 1941; rpr. 1972 by Greenwood Publishing Company, an imprint of Greenwood Publishing Group, Inc. Westport, CT. Reprinted with permission.

Alfred A. Knopf, Inc., for excerpts from *Black Majority,* by Peter H. Wood. Copyright © 1974 by Peter H. Wood. Reprinted by permission of Alfred A. Knopf, Inc. Excerpts from *From Slavery to Freedom: A History of American Negroes,* by John Hope Franklin, copyright © 1947 by Alfred A. Knopf.

Indiana University Press, for excerpts from *The New World Negro: Selected Papers in Afroamerican Studies,* by Melville J. Herskovits, ed. Frances S. Herskovits. Copyright © 1966 by Indiana University Press.

Louisiana State University Press, for excerpts from *Forty Acres and a Mule: The Freedmen's Bureau and Black Land Ownership,* by Claude F.

Oubre, copyright © 1978 by Louisiana State University Press, and *Rice and Slaves: Ethnicity and the Slave Trade in Colonial South Carolina,* by Daniel C. Littlefield, copyright © 1981 by Louisiana State University Press.

Oxford University Press, for excerpts from *Black Culture and Black Consciousness: Afro-American Folk Thought from Slavery to Freedom,* by Lawrence W. Levine, copyright © 1977 by Oxford University Press.

Sidgwick and Jackson Ltd., for excerpts from *The South as It Is, 1865–1866,* by John Richard Dennett, ed. Henry M. Christman.

I also wish to thank Horace Fraser Rudisill of the Darlington County Historical Commission for permission to use excerpts from the Journal of Evan Pugh, as typed by Mr. and Mrs. O. L. Warr, and from the John Fraser Papers; Anna S. Sherman for permission to use excerpts from her compilation of letters from Walter Gregg to Anna Parker Gregg; the South Carolina Department of Archives and History for permission to use excerpts from the Records of the Budget and Control Board, Sinking Fund Commission, Public Land Division, Duplicate Title Book B; the South Carolina State Historical Society for permission to use excerpts from the Notes of John Bennett; Special Collections, William R. Perkins Library, Duke University, for permission to use excerpts from the Plantation Book of Henry L. Pinckney, 1850–1867, and the William Law Papers, 1761–1890; Alpha Bah, J. Hazel Carter, Nelson E. Jodon, Melissa Leach, Elaine Nowick, Joseph A. Opala, Paul Richards, and Joko Sengova for permission to use excerpts from conversations and letters to the author; and Ahmadu Bello University, Timothy G. Dargan, the *Rice Journal,* Sidney Glass, Barney Mattenson, Catherine Waiters, Amelia Wallace, Matthew Williamson, and Ida Zanders for permission to use photographs, illustrations, and maps.

African Americans at Mars Bluff, South Carolina

Introduction

I WANTED EVERY word in this book to be about African Americans at Mars Bluff, South Carolina, but before I begin their story I need to explain who I am and why I was at Mars Bluff asking questions.

I was born in 1926 and grew up at Mars Bluff, a farming community in the South Carolina pine belt, about sixty miles from the coast. Tradition dictated that the lives of African Americans and European Americans were separate, but actually our lives overlapped. The first person to hold me, a newborn European American baby, was an African American woman; and it was an African American who helped me with my earliest education—"reading" the Sears Roebuck catalog.

When I was sixteen, I left Mars Bluff for college, then nursing school, a job, marriage, a move to Iowa, and five children. I returned to Mars Bluff every year for a visit with my parents—but I never intended to write about Mars Bluff or African Americans.

In the early 1970s, however, my interest was aroused when a Mars Bluff property owner wanted to tear down two irreplaceable hewn-timber houses built by African Americans about 1836. No one in the community was doing anything to prevent the proposed destruction, so I undertook the task. Each year when I went to Mars Bluff, I tried to gather information that would persuade others that the houses were worth saving.

Archie Waiters, an African American, had lived in both of the houses, so I went to see him to ask if he knew anything about how they were built. Waiters knew a lot about that subject. What was more, he knew all sorts of things about the rest of Mars Bluff—things that I was sure had never been recorded. I was amazed that one man could have so much information stored in his mind. I wanted to put it down in permanent form; so each year when I spent a couple of weeks at Mars Bluff, I would visit Waiters and ask him questions. He would hold my tape recorder and talk. He told the story of his life and his grandfather's life—the story of Mars Bluff seen from an African American point of view.

When Waiters told me that he had helped grow rice, I could not believe my ears. No European Americans at Mars Bluff knew anything about how rice had been grown in that inland area. We thought of rice as the crop of wealthy planters on irrigated plantations in the coastal region. Waiters told of African Americans growing tiny plots of rice up to the 1920s, all those years drawing on African knowledge and skills. Later on, other African Americans told me about their fathers and grandfathers growing rice. It was a fascinating example of the preservation of the African cultural heritage.

Waiters and other African Americans at Mars Bluff granted me taped interviews that transcribed into well over a thousand notebook pages altogether. They gave glimpses into the lives of African Americans in the South Carolina pine belt, a subject that had been almost completely neglected in regional histories. Just beneath the surface in those interviews was an undercurrent of an African heritage that had been retained and used, another topic that had traditionally been neglected. There was so much that needed to be put in writing for publication before it slipped into oblivion—and I was the only person with the time and interest to do the job.

I willingly admit my inadequacies for the task. My greatest shortcoming is that I am a European American, not an African American; consequently, I find painfully appropriate Lawrence Levine's quotation from an African American song:

> Got one mind for white folks to see,
> 'Nother for what I know is me:
> He don't know, he don't know my mind.[1]

Still, the African American experience at Mars Bluff deserved recording, so I gave the task my wholehearted commitment, and I look forward to a time when that experience will be described in greater depth by African Americans themselves.

I have organized this book in a way that I hope will best reveal the people of Mars Bluff and their connections with their African heritage. Part One is about the lives of African Americans in this community from the mid-eighteenth century to the mid-twentieth century. I hope that others will share my deep respect for the lives portrayed here—lives of exemplary dignity under very adverse conditions. Part Two is about African Americans' cultivation of small fields of rice. It shows a tenacious connection between the people of Mars Bluff and one aspect of their African heritage. Part Three is about a former slave named Tom Brown who used the African heritage as he faced the most critical problem of his time, the need for land.

When I wrote this book, I thought I was searching for the heritage of Mars Bluff African Americans, but I began to see the work in a different light one day when my daughter asked me a question. She wanted to know why I had called my youngest child Biddy. His real name was Andrew, but he was a baby and had hair the color of a baby chick, and at Mars Bluff baby chicks were called "biddies." *Biddy,* I learned, is a Bantu word for "small, yellow bird"; I had unknowingly given my child a Bantu name. Thus I discovered that some of the cultural heritage that I had thought belonged solely to African Americans was my cultural heritage too.

Part One: *Mars Bluff*

I
The Early Years

ALEX GREGG WALKED down a sandy road through a cotton field. He was going to see his daughter Tena Waiters; he wanted to ask a favor of her.

Gregg walked slowly, for his joints were stiff from years of labor—twenty years as a slave and fifty years as a free man. Although age had slowed his pace, still his stocky body was erect and he held his head in a proud way.

He found Tena and her husband, Otis, at home with their three small children, and he made his request: "Emma needs a baby to pacify her."

Emma was Gregg's wife. She had been a very active person, but now she was in poor health and housebound. The couple's twenty-three children were grown and working, so they were no company for their mother. She wanted a baby to keep her company, so Tena and Otis gave their three-year-old son, Archie, to his grandparents.[1]

Gregg took Archie home with him and kept him throughout his childhood. Archie worshiped his grandfather and called him "Pa." In the evening, when the old man would reminisce, Archie would sit on the porch steps or by the fireplace and listen. Even after he was grown and no longer lived with his grandfather, Waiters visited him regularly. They would sit and talk, and Gregg would retell his stories of long ago. By the time Gregg died in 1938, Waiters had learned the stories so well that years later he could retell them, just as if Alex Gregg were still alive and reminiscing.

Gregg remembered slavery, for he was born a slave at Mars Bluff, South Carolina, in 1844; and his memory stretched even farther back in

Fig. 1 Alex Gregg and his second wife, Florence Henderson Gregg, probably about 1930.

Photograph courtesy of Catherine Waiters.

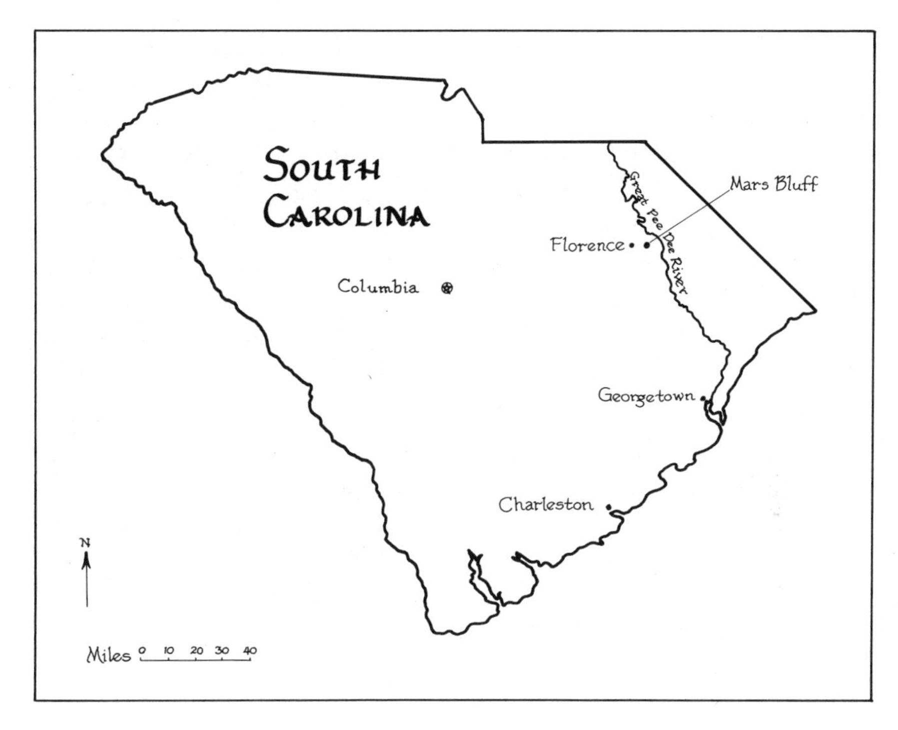

Map 1 Location of Mars Bluff. *Map by Dinah Bervin Kerksieck.*

the stories he had heard the older people tell.[2] When Gregg described how Africans had come to South Carolina, he told the story as he had heard it told when he was a boy in the 1850s. When Waiters repeated the story, he told it as his grandfather had told it to him when he was a little boy.

> Pa say they come from Africa here.
> I say, "Well how you get cross the water?"
> He say, "Come by boat, boy."
> I say, "Come by boat. How long it take you?"
> "Six months and a year."
> "How many come on the boat with you?"
> "A heap of them," he say. "Heap of them. Don't know how many come over."
> I said, "When you landed, how you do it?"
> Ship couldn't get close. Got off the ship on to a flat built over the water. The ship pull up there and that when they un-

> load them. . . . Long walk on a flat and they come up in a building. . . .
>
> Ship come in at Georgetown. That's where they unload at. . . . And men out there picking out who they want to get. Heap of people be there buying. "Come over here." "Come over here." "Come over here." Just like that and buy them out.
>
> He say about three shipload of them people come over here that he know of.
>
> And I say, "How long the ship was?"
>
> He say, "Well, it take most a day to unload it because upstairs and downstairs the ship."
>
> [Waiters stopped talking like his grandfather.]
>
> Georgetown dock was where they unload them, and the mule and wagon haul them here [Mars Bluff]. Some of them drive the mule and wagon down there—bring them back.
>
> When they get where they going, Pa say the boss will tell him to call him "Boss" and to call the boss's wife "Missy." Call him "Boss" and call the wife "Missy." That what he tell me. That the way it is, man.[3]

The Africans' arrival in Georgetown as described by Gregg was not the typical arrival from Africa. Most Africans entered South Carolina through the port of Charleston.[4] Also, many were not sold immediately off the ship, but were held and made more presentable before sale. Gregg's story seemed to be of a time when the demand for slaves was so great that buyers competed for them as they disembarked.

Although Gregg's story is atypical, it is important, for it is the only one of its kind known. All other stories about the arrival of Africans at Mars Bluff have been lost. No one recorded the lives of Mars Bluff slaves in the early days of the settlement.

On the other hand, the lives of European American settlers at Mars Bluff were well recorded. In 1730 there were only a few scattered settlers in the area. A few years later, the government adopted a policy to attract people to this wild backcountry, offering fifty acres of land for every adult in a household. Settlers arrived both from the British Isles and from colonies to the north to claim land. Most were English, Scotch, Irish, or Welsh. A group of Welsh people came from Pennsylvania to settle north of Mars Bluff, while a group of Scotch-Irish settled to the south in Williamsburg. Some of the Scotch-Irish moved on to Mars Bluff in search of more land.[5] The stories of all of these European Americans were written. Only the African American part of the Mars Bluff story was missing.

In fact, the African American history of most communities was missing and probably would never be recovered. At Mars Bluff, however, there seemed to be a chance that part of the story might be found. Waiters knew so much that his grandfather had told him; perhaps by piecing together Gregg's recollections and information from the other meager sources available the story of African American life at Mars Bluff might be reconstructed. That hope was worth pursuing. So much was lost—any reclamation would be worthwhile. I resolved that I would try to reconstruct the story of African Americans at Mars Bluff.

Mars Bluff was a community of farms. It lay sixty miles from the coast in the South Carolina pine belt, a sandy, forested region that covered most of the eastern part of the state. Between the pine belt and the sea lay a narrow strip of land known as the coastal region. Life for slaves in those two regions differed greatly.

There were large rice plantations in the coastal region, and most of its inhabitants were African Americans. They had very little contact with European Americans. Sometimes, on a large plantation, there were scores of slaves and only one European American, the overseer.[6] In that isolated environment, the enslaved workers were able to retain their African customs; they even developed a language of their own, called Gullah. It enabled Africans who spoke many different languages to talk to one another.[7]

On the other hand, in the pine belt the farms were much smaller, and the number of African Americans and European Americans was about equal. Most slaves were in close contact with European Americans. Consequently, pine belt slaves spoke a dialect that was much like standard English, and their African customs were blended with European customs.

Because the people who spoke Gullah were unique, early scholars focused on them. Pine belt African Americans were ignored. However, even now, if the scattered facts that can still be found could be pieced

Fig. 2 Archie Waiters with tape recorder. Waiters told stories of South Carolina—stories filled with hints of Africa.

Photograph © 1987 Sidney Glass.

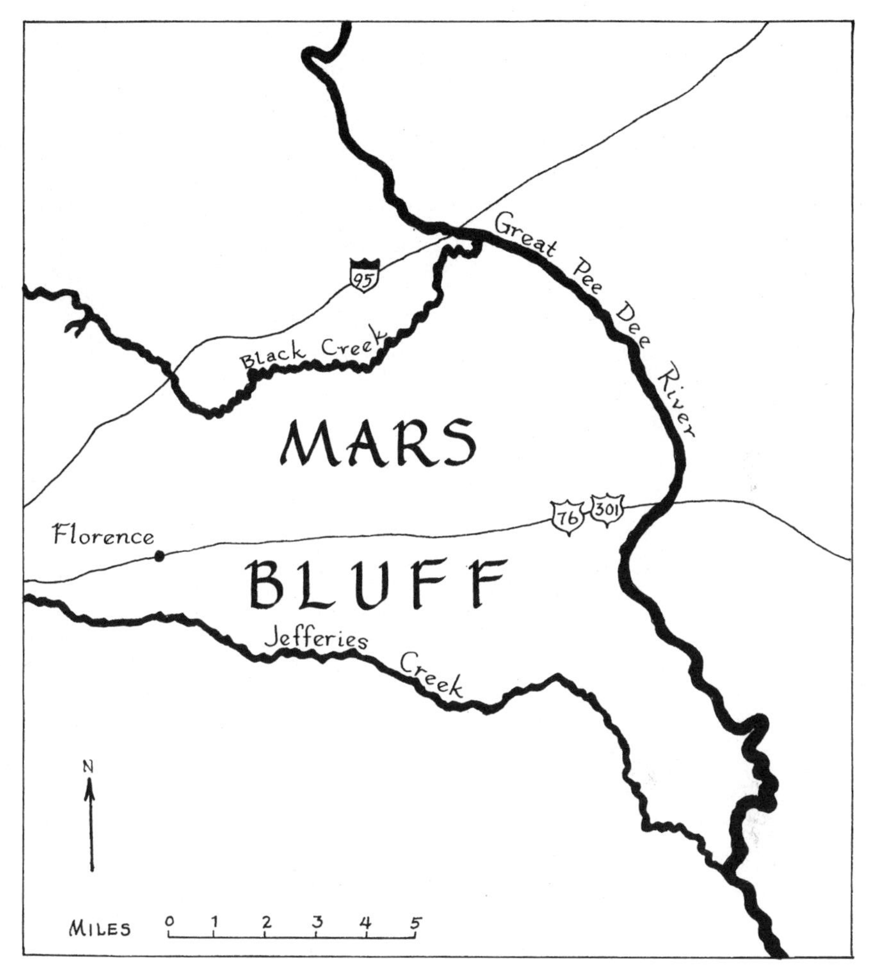

Map 2 A sprawling farming community, Mars Bluff is about eight miles across in both directions and consists largely of pine and hardwood forests on lowland that is too wet to farm. In the nineteenth century the common crops on the high land were cotton, corn, and peas.

Map by Dinah Bervin Kerksieck.

together, part of their story might be told. Although little is known about any one community, if every pine belt community around Mars Bluff was searched, perhaps enough information might be found to draw a picture of the African American experience in the area. That larger picture would give a glimpse of what life must have been like for African Americans at Mars Bluff.

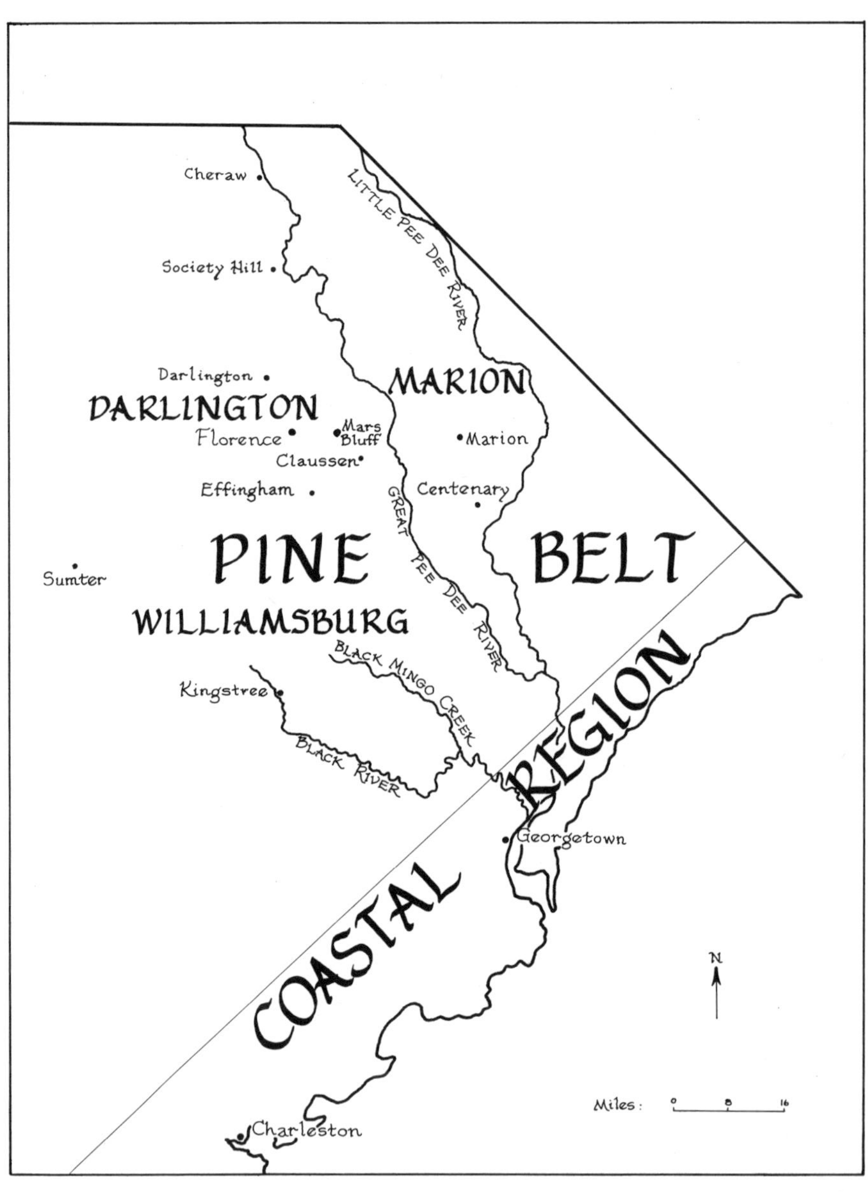

Map 3 Northeastern South Carolina pine belt and coastal region in the mid-nineteenth century, showing places referred to in the text. Names of counties are in boldface lettering.

Map by Dinah Bervin Kerksieck.

The first question that cried out for an answer was, Who were the Mars Bluff African Americans? Where in Africa had they originated?

Daniel C. Littlefield supplied a basic answer to that question. He studied eighteenth-century records of ships that brought Africans to Charleston and concluded that 43 percent came from the rice-growing region on the Windward Coast, and 40 percent from the Congo-Angola region. According to those figures, there was a good chance that many of the people at Mars Bluff had ancestors from one of those two regions.[8]

Littlefield's figures were helpful, but they did not pertain specifically to Mars Bluff. Would it be possible to find more precise information about the African provenance of individuals who had come to Mars Bluff or nearby communities?

Such information was seldom available. Rarely did one see an African country or ethnic group named in relation to an individual African American. One former Sumter County slave, Jacob Stroyer, supplied the name of a country when he wrote: "My father was born in Sierra Leone, Africa. Of his parents and his brothers and sisters, I know nothing; I only remember that it was said that his father's name was Moncoso, and his mother's Mongomo."[9]

It was unusual for a slave to have the means to leave a written record, so it is not surprising that not a single record like Stroyer's exists for Mars Bluff. Furthermore, no one I spoke with at Mars Bluff could recall where their people had originated in Africa—so many generations had come and gone. It seemed that the provenance of Mars Bluff African Americans would remain a complete mystery.

Then Archie Waiters accomplished the impossible. He supplied clues that connected some Mars Bluff people to specific places in Africa. Were these connections circuitous and meaningless—or was it possible that these people had forebears who came from these places in Africa?

Waiters gave the first clue when he was describing an African American cemetery at Mars Bluff. He said: "The biggest tombstone was to Mariah Malinka. My grandfather told me that she was one of the first people buried there."[10]

The name seemed strange; I was not sure that I had heard it correctly, so I asked Waiters to repeat it.

"Mariah Malinka," he said.

I wondered about the name. I thought that it must be a Scotch "Mc" name that he was mispronouncing. Mars Bluff was settled by the Scotch and had innumerable names with a "Mc" prefix but no names that sounded like "Malinka."

I asked Waiters, "Say it again?"

"Mariah Malinka."

Here was a substantial clue that I missed completely. Its significance only dawned later when I was reading about the ancient African kingdom of Mali, inhabited by the Malinke.[11] Mariah Malinka. Her last name was the name of an African ethnic group.

In fact, it was a highly respected name, for the Malinke had played an important role in the development of Mali, a powerful empire that dominated most of West Africa in the thirteenth and fourteenth centuries. Controlling trade between North Africa and the gold mines to the south, Mali became enormously wealthy. Although it drew attention for its efficient government and its scholars, it was most widely known for its king who journeyed to Mecca with eight thousand retainers and unbelievable amounts of gold.[12]

The kingdom of Mali was built around inland trade cities on the Niger River. During the centuries when the kingdom was at its height, the Malinke expanded the empire to the south and west. Many of them were living in the region around the Gambia River when the Portuguese slave traders arrived there. These people who had contributed so much to the development of the great Mali empire became one of the ethnic groups considered most desirable by South Carolina slave merchants and planters. Michael Mullin wrote of the preference that South Carolinians showed for slaves from Senegambia—and for the Malinke.[13] So ship captains tried to fill their holds with the Malinke because they brought good prices in South Carolina.

It is possible that Mariah Malinka had been captured in Africa and had made the terrible journey across the Atlantic, or she may have been born in America, a child of a Malinke from Africa. But it is also possible that she had little connection with the Malinke, that the name was arbitrarily chosen for her by someone who had heard the name and liked it. No one will ever know how close or how remote Mariah Malinka's connection was to the Malinke in Africa. Still, that name is the only evidence of its sort found in the search for the African provenance of Mars Bluff African Americans. Mariah Malinka was the

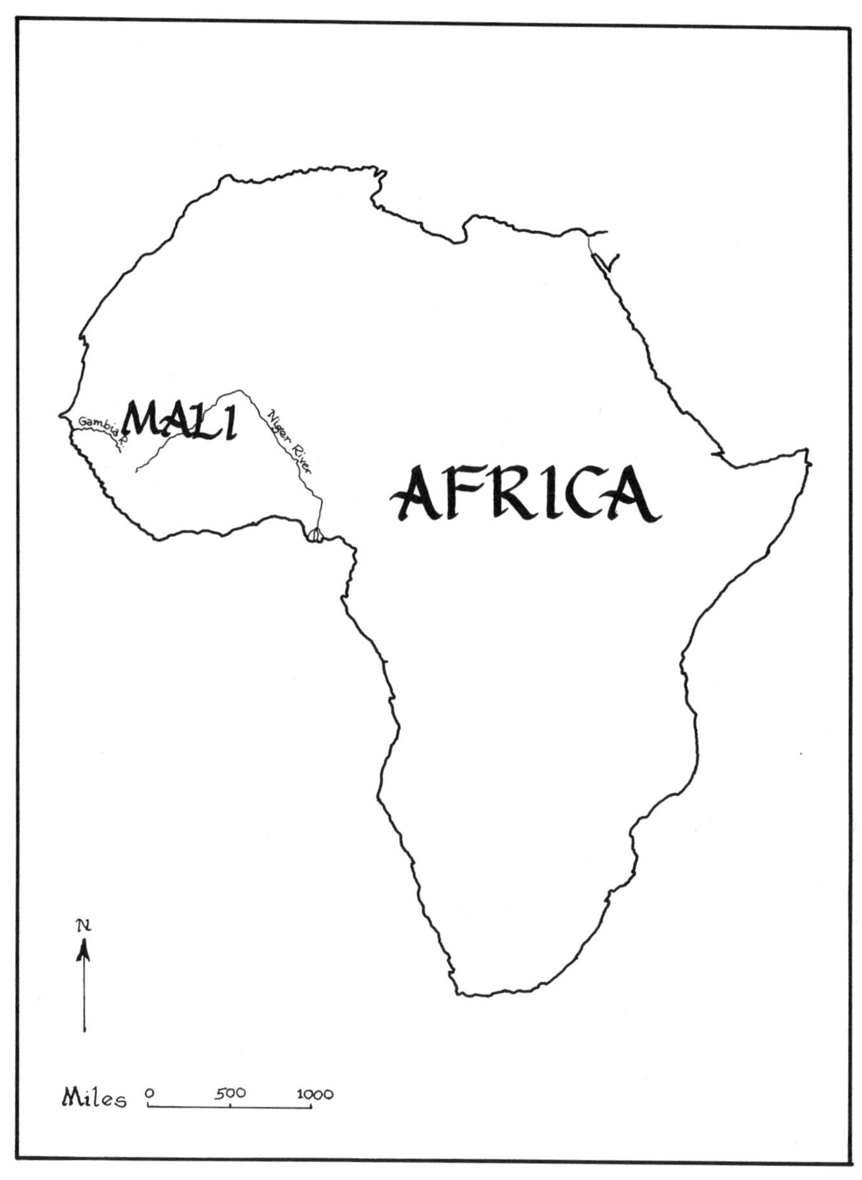

Map 4 The kingdom of Mali during the thirteenth and fourteenth centuries.
Map by Dinah Bervin Kerksieck.

only person who left a clue that pointed directly to a specific ethnic group.

Waiters supplied another clue. It was the expression "Great Da!" Frequently, Waiters said "Great Da!" when he heard some fact that was remarkable or when he saw something that was unusually large or strange.

Waiters told one story in which his grandfather Gregg used the expression. When Waiters was a boy, his grandfather had some hogs that he was trying to fatten. He told Waiters to give each of them one ear of corn a day. Knowing that the hogs would never gain weight on that diet, Waiters developed a daily feeding schedule of his own.

"I'd get a basket of corn," said Waiters, "and carry it way back on the back side where Pa wouldn't go."

When the hogs were well fattened, Waiters' grandfather looked at them in amazement and said, "Great Da! They sure is coming out."[14]

For years I had heard Waiters say "Great Da!" and had always assumed that *Da* was a meaningless syllable, but then I happened upon a book by the anthropologist Melville Herskovits. It was about the religion of the people of Dahomey, a small country on the west coast of Africa, now called Benin. The Dahomean religion was so complex that a whole book was needed just to outline it. In the pages that listed the names of the gods, one name caught my eye—*Da*. Da was a Dahomean god.

Waiters' "Great Da!" had profound meaning. It was part of a complex African religion.

Da himself is very complex. He has numerous aspects because he is found in all things that are flexible and sinuous. He is present in the snake that supports the earth, the rainbow-serpent, the spirits of the ancestors, and the umbilical cord of each person.

Da is an important god because it is he who brings good fortune and bad fortune. Herskovits wrote that Da is a god to be feared, not loved or esteemed. "Da is a thief and a fickle one, for what he gives to one he takes away from another. . . . One must be careful with Da, for he doesn't forgive easily."[15]

So that was who Waiters' "Great Da!" referred to. Da wasn't just any god; he was the god who brought good and bad fortune. Forgetting him was a dangerous thing to do. Perhaps that is one reason he had been remembered through so many generations in the New World.[16]

Gregg and Waiters' use of the word *Da* could be a clue that one of Gregg's ancestors came from Dahomey.[17] Such speculations are tentative at best. They lead to suppositions—not facts—about African connections. Still, each clue is as exciting in its own small way as was Alex Haley's search for Kunta Kinte's home in Africa.[18]

Waiters' memories of African words like *Malinka* and *Da* suggested places in Africa where the forebears of some Mars Bluff African Americans may have originated. Waiters had no stories, however, about the next phase in the history of African Americans—the crossing of the Atlantic. If Alex Gregg knew any of those stories, he never revealed them to Waiters.

Another resident of Mars Bluff did recall hearing people talk of being taken from Africa. Charlie Grant, who was about the age of Alex Gregg, was interviewed by the Federal Writers' Project in 1937. He said:

> I hear dem tell dat my grandparents come from Africa. Dey fooled dem to come or I calls it foolin dem. De peoples go to Africa en when dey go to dock, dey blow whistle en de peoples come from all over de country to see what it was. Dey fool dem on de vessel en give dem something to eat. Shut dem up en don' let dem get out. Some of dem jump over board en try to get home, but dey couldn' swim en go down. Lots of dem still lost down dere in de sea.[19]

And then there was the journey from Africa to the New World—and on to Mars Bluff. Most Africans had probably taken generations to make that trip. They might have gone first to the West Indies, or to Virginia, or to the South Carolina coastal region, or to other pine belt communities. They may have lived in one of those places for years—or for several generations. The stories were all lost, except for tiny bits of information that could be pieced together.

One clue suggested that some few may have come directly from Africa to Mars Bluff. That clue came from Black Mingo, a community in the Williamsburg area south of Mars Bluff. A Charleston merchant wrote a letter to a settler in Black Mingo about his request to buy newly arrived Africans:

> 18 May 1764
>
> Dear Sir,
>
> I thought there would be a better opportunity of purchasing Negroes for you some time after the last Sale that was made & therefore deferred that business.
>
> No Guinea Ship arrived since but you may rest assured that I shall be mindful of your order whenever there appears to be a proper time to execute it, & I remain, Dear Sir.[20]

Africans, direct from Africa, were going to Black Mingo. This suggests that they were probably also going to Mars Bluff, for the two communities had much in common. The areas were settled at approximately the same time by the Scotch-Irish and had similar environments and farming practices. In fact, people from Black Mingo and other Williamsburg settlements moved to Mars Bluff, so some Africans may have lived a short time in the Williamsburg area and been required to move to Mars Bluff when their owners moved in search of more land.

Because most slave ships used the Charleston port, Africans would have arrived at Charleston and stayed there while awaiting sale. Gregg told a story he had heard from older slaves that sounded as if they might be recalling their first days in Charleston, when they were newly arrived from Africa.[21] Waiters repeated the story.

> It have to be a big, large building to hold about forty or fifty head of people. Like white man got his house here. Got us house setting off about an acre or two acres from his house. And a fence go round it, go all the way around that thing. . . . Fence ten or fifteen feet high, couldn't climb over. If you come back, got to come by him. Couldn't go out no other way—come through the gate by him. And you got to hail them before you come through or the dogs will eat you up.

Big old place, and straw in there, and bag and sheet spread over them. Women sleep here. Men sleep here. Girls sleep here. Boys sleep here. Nothing but straw on the floor. Chimney in the middle—big old thing burn four-foot wood—back off in the middle of the place. Fire it once a day. Get them old four-foot logs. That thing would burn all day and night.

They hold a light up to your window and jerk it down like that. You wake up and see that thing. They jerk it right back down.

"You see that?"

"You see that?"

Those people be running over one another. They didn't know what it was. Didn't know what it was—the man doing that—hold up to this window, get you running. Go to the next window. All of them start to running. That was enough to scare you. Didn't know what it was.

Start to running, running over one another. . . . Some of them get their arm break, leg break, where you get knocked down. People run over and stepping all over you.

They soon stopped that.

The slaves who told this story may have been recalling a barracoon in Charleston.[22]

One record exists about the next stage of the journey, traveling from Charleston to the pine belt. After a slave was sold in Charleston to someone in the Mars Bluff vicinity, there was an overland trip of well over a hundred miles—much longer by water. One such trip was mentioned in the journal of the Reverend Evan Pugh, a farmer and Baptist preacher in the Cheraw District, a settlement north of Mars Bluff. Regrettably, the journal entries give no information about the slave involved—whether he was newly arrived from Africa or was being resold after having lived in South Carolina or some other colony.

In June, 1765, Pugh wrote:

Fig. 3 Former site of a Charleston barracoon, according to a flyer from the Old Slave Mart Museum. It stood at 15–17 Queen Street and was called Ryan's Jail or Barracoon. Many of the people who sold slaves had places of business in this part of Charleston.

> Thurs. 6, I got to Towne [Charleston]. Bought a negro boy. . . .
>
> Fri. 7, In Towne walked about. Sent the negro out with Able Wilds [a European American from the Cheraw District].
>
> Sun. 9, Felt poorly. Mr. Hart preached.
>
> Mon. 10, I sot off for home. Lodged at Capt. Lashoes' plantation.
>
> Tues. 11, Came to Potts'.
>
> Wed. 12, Over took the negroes. Staid with them the night in the woods.[23]

Although Pugh revealed nothing about the African American, he did say something about local conditions in his entry for June 12. The pine belt was rough frontier country where conditions were harsh for everyone. The newly purchased slave and Pugh slept in the woods together.

That beginning was an indication of the close association that the young slave would have with European Americans. He, like many other African Americans in the pine belt, would live on a small farm where he would be in daily contact with European Americans.

One wonders about the trip that young African American made from Charleston to the pine belt: what sort of security measures would Able Wilds have taken to keep him from running away at night while they were sleeping? Charles Ball gave an account of the absence of such measures in South Carolina. Ball, who walked from Maryland in irons, said that once he and his fellow slaves reached South Carolina, their irons were taken off. All that was needed was for the slave trader to point out that "it was impossible for them to escape without being taken up and sent back."[24]

When Ball was sold in Columbia, his new owner left him waiting outside a tavern for hours with no one watching him. When the owner emerged from the tavern, it was too dark to see Ball, and the European American said to his companions, "I bought a Negro this evening,—I wonder where he is." Ball presented himself from the darkness with the reply, "Here, master."[25]

Perhaps such lax security was the norm in South Carolina. Perhaps all that was necessary was for the slave to know that if he escaped he would be recaptured—and would be punished very severely.

One would expect that many slaves in the Mars Bluff area would have come from Georgetown rice plantations, for boats on the Great Pee Dee River moved between the two regions regularly.[26] But Charles Joyner, who was most knowledgeable about slavery in the Georgetown rice region, believed that very few slaves were sold from that region to places like Mars Bluff; planters rarely sold slaves from rice plantations except when estates changed hands. Also, the increase in rice production over the years would indicate that the rice plantation owners were not likely to be selling African Americans inland.[27]

Still, there is evidence that some African Americans in the Mars Bluff area had previously lived in the coastal region. Chapman Milling reported that at Mechanicsville, a community a few miles north of Mars Bluff, "Many spoke with a Gullah accent, having been brought up the river from Charleston and Georgetown."[28] Even today, some of the same words are used in Mars Bluff as in the Gullah region.[29]

Waiters made one remark about Georgetown that seemed inexplicable. He said: "When they [landowners at Mars Bluff] want to take those people [slaves] what working for them back down there . . . get in touch with their family overseas . . . over in Africa . . . go to Georgetown to find out about the people down there. You get in touch with them."[30]

What was the meaning of that statement? It seemed impossible that Mars Bluff slaves would have expected to make contact with their families in Africa by going to Georgetown—but actually there would have been a possibility. In the Georgetown area, where the concentration of Africans was so great, it would have been possible for a new arrival from the same place in Africa to bring news of a family in Africa. In fact, any person of the same ethnic group was considered "their people." So it was reasonable for Mars Bluff slaves to hope that a trip to Georgetown would help them get in touch with "their people" in Africa. They might not be able to send messages back to their homeland, but they could get news of a general sort about their ethnic group in Africa.

No doubt very few African Americans at Mars Bluff were African born. Still, that statement from Waiters, that slaves at Mars Bluff wanted news from their families in Africa, seemed to suggest that some people at Mars Bluff were African born.

While people at Mars Bluff were yearning for news of their families in Africa, in Africa, according to Herskovits, there was a similar yearning for contact with family members in America. The people of Dahomey prayed that the white strangers who carried their relatives off to America would bring them news of those relatives. They also prayed that "the Americans . . . bring the cloths and the rum made by our kinsmen who are there, for these will permit us to smell their presence."[31]

Probably quite a few pine belt African Americans came south with their owners who were leaving North Carolina, Virginia, and Maryland in search of new land. A 1745 record shows a Virginian moving into the Cheraw District, applying for the land to which he was entitled because he was bringing twenty-five settlers, "doubtless most of them slaves." African Americans would continue to come into South Carolina from other states; however, in 1787, South Carolina prohibited the importation of slaves from Africa. When the cotton gin was invented six years later, communities like Mars Bluff realized that fortunes could be made on cotton. All that was needed was more slaves to clear more land and cultivate more cotton. Wanting to import greater numbers of Africans, cotton planters succeeded in reopening the slave trade for four years, from 1804 through 1807. Then the United States government stopped all African slave trade. No slaves would come to the pine belt from Africa after 1807.[32]

At this time, land in Virginia was becoming too exhausted to use, and the African American population there was increasing. Thus many Virginia slaves were sold to the south and west. Some were brought to the pine belt by slave traders. Jacob Stroyer, who had been a slave in Sumter County, wrote that a mulatto woman named Moriah was brought from Virginia by traders. "She was very beautiful, and wherever she was sold her mistresses became jealous of her, so she changed owners very often. She was finally sold to Boney Young who had no wife."[33]

Ball gave a graphic description of the journey that slaves endured when they were brought south from the Middle Atlantic states. Some African Americans in the pine belt probably had made journeys similar to Ball's, and at about the same time—shortly after the invention of the cotton gin—for there was a great demand for slaves in the pine belt at that time.

Ball, a Maryland slave, was sold to a slave trader. He was forced to leave immediately with fifty-one other slaves, given no chance to tell his wife and children good-bye. Of his companions on the journey Ball said:

> Thirty-two of these were men, and nineteen were women. The women were merely tied together with a rope, about the size of a bed cord, which was tied like a halter round the neck of each; but the men, of whom I was the stoutest and strongest, were very differently caparisoned. A strong iron collar was closely fitted by means of a padlock round each of our necks. A chain of iron, about a hundred feet in length, was passed through the hasp of each padlock, except at the two ends, where the hasps of the padlocks passed through a link of the chain. In addition to this, we were handcuffed in pairs, with iron staples and bolts,

> with a short chain, about a foot long, uniting the handcuffs and their wearers in pairs. In this manner, we were chained alternately by the right and left hand; and the poor man, to whom I was thus ironed, wept like an infant when the blacksmith, with his heavy hammer, fastened the ends of the bolts that kept the staples from slipping from our arms.[34]

They walked for a month, all the way to South Carolina, chained together in that way. Of his arrival there, Ball said:

> I shall never forget the sensations which I experienced this evening, on finding myself in chains in the state of South Carolina. From my earliest recollections, the name of South Carolina had been little less terrible to me than that of a bottomless pit. In Maryland, it had always been the practice of masters and mistresses, who wished to terrify their slaves, to threaten to sell them to South Carolina. . . . I seriously meditated on self-destruction, and had I been at liberty to get a rope, I believe I should have hanged myself.[35]

No one knows how many African Americans may have made a trip similar to Ball's from the Middle Atlantic states to Mars Bluff.

One of the few clues I found about how African Americans came to Mars Bluff involved the Fleming family. They reversed the direction that most people were moving.

Early in the nineteenth century, thousands of people moved from East Coast states to Alabama, Mississippi, Louisiana, and Texas. Settlers went in search of more fertile land for new cotton fields.[36] Slaves were brought to clear the land and grow the cotton—some taken by their owners and some by slave traders.

When I was a child, if the name *Fleming* was mentioned in my grandmother's presence, she would always say: "The Flemings are nice Negroes. They came from Louisiana."[37] From Louisiana to South Carolina—that was opposite to the direction that most people were moving. Unfortunately, I never asked for an explanation of the Flemings' route, but I am fairly certain that it was determined in this way. When my grandmother's grandmother was a girl, her parents, who had migrated from South Carolina to Louisiana, died. The orphan girl was sent back to Society Hill, South Carolina, to be raised by relatives. Years later, after she was married, she moved from Society Hill to Mars Bluff.

I would surmise that the Flemings had been purchased in Louisiana by the girl's parents. When the parents died, possession of the Flemings passed to their daughter. Consequently, each time she moved, the Flemings were obliged to move with her.[38]

The story of the Flemings raised more questions than it answered. Had they originally entered the country through an Atlantic or a Gulf port? Many of the slaves in Louisiana had been brought from the Middle Atlantic states by slave traders. Had the Flemings been among the thousands who made that trip?

Carter G. Woodson quoted from a man who had seen the suffering that slaves endured while they walked from the Middle Atlantic states to new cotton lands in the Gulf states:

> The hardships these Negroes go through who are attached to one of these migrant parties baffles description. They trudge on foot all day through mud and thicket without rest or respite. Thousands of miles are traversed by these weary wayfarers . . . urged on by whip. . . . I have never passed them staggering along in the rear of the wagons at the close of a long day's march, the weakest furthest in the rear, the strongest already utterly spent, without wondering how Christendom . . . can look so calmly on at so foul and monstrous a wrong as this American slavery.[39]

Had the Flemings made that terrible journey? No one will ever know.

Unfortunately, there are no records like Ball's to tell what African Americans were thinking and feeling in their early years at Mars Bluff. Still, by piecing together the clues that can be found, a picture begins to emerge—a picture of where in Africa Mars Bluff African Americans may have originated and how they were brought to Mars Bluff.

2
The Last Years of Slavery

"ALL OF THEM come from Africa. Everyone of them come from Africa. . . . The name they give them, they ain't named. Like my granddaddy . . . he named Rooster. . . . They call him Alex Gregg."[1] Archie Waiters was reminiscing about things his grandfather had said.

Waiters remembered so much that his grandfather had told him. Those recollections offered a chance to hear in an African American's own words what slavery at Mars Bluff had been like. In addition, a Federal Writers' Project interviewer in 1937 had recorded the experiences of several former slaves who had lived in or near Mars Bluff.[2] If those stories were pieced together, Alex Gregg and other formerly enslaved African Americans could tell in their own words what life had been like at Mars Bluff in the last years of slavery.

Gregg told Waiters about where the slaves lived in the early days, when all South Carolina settlements were oriented toward the rivers and creeks. Waiters said: "My granddaddy was born down there . . . in houses most down to Jefferies Creek. Was about ten log houses back in there. They moved back up this a way because the water would override the bank every time the creek would rise. Build ten more houses out on the field side, and all of them come out there."[3]

Ten houses. That was quite a few for Mars Bluff. That meant, by pine belt standards, that Alex Gregg lived on a fair-sized farm where there were more slaves than usual.[4] Although there had been few African Americans in the early days of the settlement, every year following the invention of the cotton gin in 1793 Mars Bluff planters had tried to acquire more land and slaves so that they could raise more cotton.[5]

Waiters told about the kind of work that his grandfather had done when he was a slave.

> Pa say cleaning land, that was his work. See, back there then Pa say they ain't had enough land clean up to do nothing with. Cutting those trees. Rolling those logs out the way. Set them on fire. Digging those stumps.
>
> Had other people to done the other little work—other people plowing. Just like I'm a little man; you's a little woman; us do the plowing, make a little crop. But them big people, they clean up them logs and roll them together. Wasn't nothing but clean up land back then. Women worked too. They worked just like the men do. Older woman, what done worked down, do most tending the babies.[6]

These words give a picture of one African American's life—everyday he faced the herculean task of turning a forest into cleared land. The traditional picture showing many slaves working together in cleared fields gives no idea of the tremendous labor they had exerted to clear those fields.

Also, it is difficult to believe what Gregg said—that women worked just as men did in the backbreaking labor of land clearing. Peter Wood, however, confirms Gregg's statement. Wood wrote that frontier farmers in Carolina used women in land clearing and expected them to perform tasks comparable to those of the men.[7]

After the enslaved workers finished their day's work, they went to a central kitchen, where they all ate together out of troughs.

> Pa say they go out and work and just come in and wash up and eat. They had a cook. Just like you cook. . . . You fix it on the table in a big old trough—like the milk and mush and stuff like that. . . . Just take that buttermilk and stuff and pour it right on down in the trough. And got them big old wood spoon. . . . Everybody eat out that trough. They get cabbage or collards or peas, something like that; that be on table on little wooden plate.
>
> For breakfast, they ate grits and butter and a big piece of corn bread. For dinner, they ate mush and milk, sometime eat collard and peas and stuff like that. Eat a light supper, sometimes mush and milk, sometimes eat rice and butter and a piece

> of corn bread. After supper at night everybody would help shell the peas or beans for the next day.

Mush, grits, or cornbread was eaten at every meal. All three were made of corn; the mush was just finer-ground corn than the grits.[8]

It was surprising to hear Waiters say that the cook poured the mush and milk into a trough. The slaves ate from a trough. Waiters described the troughs as similar to wooden containers used for making bread, only longer so that a number of people could eat from one trough at the same time. He said that he had seen a trough that his grandfather's brother had kept.[9]

There was another report about slaves eating from wooden trays. Charles Ball told about a meal he and his fellow slaves were given shortly after they reached South Carolina. He said, "At supper this night, we had corn mush in large wooden trays, with melted lard to dip the mush in before eating it." Was it significant that Ball mentioned the trays? It could mean that he considered eating from them unusual—a practice he had not seen until he reached South Carolina.[10]

Later, when Ball was living on the Congaree River, east of Columbia, he said that he had worked on Sunday and earned money to buy tools to carry on his trade, carpentry. Then he remarked: "I occupied my leisure hours for several months after this, in making wooden trays, and such other wooden vessels as were most in demand. These I traded off, in part, to a storekeeper."[11] Ball did not explain why wooden trays were in demand. Perhaps they were used routinely by slave owners.

Years later, Gregg told his grandson about courting Emma Pigiott while they were slaves living on different farms.[12] Waiters recalled:

> My grandmother [Emma Pigiott Gregg] house used to been back up in the woods next to the creek—skip this place here and on that place [two farms west of the farm where Gregg lived].[13] Like Pa be off on a Sunday, he go to his boss and get a note to go over there see her. Pa say she was shape up like a Coca-Cola bottle and look like a peach honey.
>
> I say, "Good Pa."
>
> Pa say, "I love that woman."

I say "You love her?"

"Yeah, I love her."

I say, "I love her too." . . .

Get a note to go over there to see her. . . . Visiting. That the way you had to do. They didn't go out there on your own now. Just get a note to go these places, then go in peace—wouldn't get beat. . . . If you ain't had that note, when you come back. . . . boss make those other fellows catch him and like to beat him to death. . . .

Us have to do what Boss say. . . . Beat him until Boss tell them to quit, "Don't beat him no more."

Pa says he sees two get beat like that. Says he ain't never tried it [going off without a note].[14]

Alex Gregg had told something of what life was like for a man whose work was clearing land. Another part of the story was told by Sylvia Cannon in a Federal Writers' Project interview in 1937. Mrs. Cannon was a former slave who had lived in Effingham, just a few miles southwest of Mars Bluff. She described her life as a girl, when her work was the care of her owner's children.

Father en Mother belong to de old Bill Greggs and dat wha' Miss Earlie Hatchel buy me from. After dat, I didn' live wid my parents any more, but went back to see dem every two weeks. Got a note en go on a Sunday evenin en come back to Miss Hatchel on Monday. Miss Hatchel want a nurse and dat howcome she buy me. I remembers Miss Hatchel puttin de baby in my lap en tell me don' drop him. Didn' have to do no work much in dem days, but dey didn' allow me to play none neither. When de baby sleep, I sweep de yard en work de garden en pick seed out de cotton to spin. Nursed little while for Miss Hatchel en den get free.

I see em sell plenty colored peoples away in dem days cause dat de way white folks made heap of dey money. Coase dey ain' never tell us how much dey sell em for.[15] Just stand em up on a block bout three feet high en a speculator bid em off just like dey was horses. Dem what was bid off didn' never say nothin neither. Don' know who bought my brothers, George en Earl. [She cried after this statement.] I see em sell some slaves twice

DUTIES

OF

CHRISTIAN MASTERS.

BY

H. N. M'TYEIRE, D.D.

EDITED BY THOMAS O. SUMMERS, D.D.

Nashville, Tenn.:
SOUTHERN METHODIST PUBLISHING HOUSE.
1859.

Fig. 4 *Duties of Christian Masters,* published in 1859. Even in a book with such a promising title, corporal punishment was sanctioned. The book belonged to a landowner at Mars Bluff who lived within a few miles of Alex Gregg's boyhood home.

before I was sold en I see de slaves when dey be traveling like hogs to Darlington. . . .

Many a time my Missus go work in de field en let me mind de chillun.

We live in de quarter bout one-half mile from de white folks house in a one room pole house what was daubed wid dirt. Dere was bout 20 other colored people house dere in de quarter dat was close together en far apart too. De ground been us floor en us fireplace been down on de ground. Take sticks en make chimney cause dere won' no bricks en won' no saw mills to make lumber when I come along. Oh, my white folks live in a pole house daubed wid dirt too. Us had some kind of homemade bedstead wid pine straw bed what to sleep on in dem days. Sew croaker sack together en stuff em wid pine straw. Dat how dey make dey mattress. . . .

De white folks didn' never help none of we black people to read en write no time. Dey learn de yellow chillun, but if dey catch we black chillun wid a book, dey nearly bout kill us.

I go to Church wid my white folks, but dey never have no church like dey have dese days. De bush was dey shelter en when it rain, dey meet round from one house to another. Ride to church in de ox cart cause I had to carry de baby everywhe' I go. White folks didn't have no horse den.[16]

Mrs. Cannon supplied a picture not only of her own life but of her owner's as well. He certainly was not the traditional picture of a slave owner—a gentleman in a white suit. He lived in a pole house, had no horse, and used an oxcart to take his family to church. His wife sometimes worked in the field. At Mars Bluff, many slaves probably had owners like him.

Also insightful was the remark, "If dey catch we black chillun wid a book, dey nearly bout kill us." Not only were the children not taught to read, they were punished if caught with a book. That practice was confirmed by another former slave who lived on the north side of Mars Bluff on one of the more prosperous farms. Charlie Grant said: "Dey catch nigger wid book, dey ax you what dat you got dere en whe' you get it from. Tell you bring it here en den dey carry you to de whippin post for dat."[17]

Mrs. Cannon, in telling about the sale of slaves, had spoken of another subject on which Alex Gregg had been silent. It is possible that there were no sales on the farm where he lived. Even Mrs. Cannon had been more fortunate than most in having been sold to a nearby owner who allowed her to visit her parents. Her brothers' fate, being carried away and never heard from again, was more common.

Former slave Lizzie Davis, who lived in Centenary, in lower Marion County, told a similar story: "I remember, I hear my father tell about dat his mammy was sold right here to dis courthouse, on dat big public square up dere, en say dat de man set her up in de wagon en took her to Georgetown wid him. Sold her right dere on de block. Oh, I hear dem talkin bout de sellin block plenty times. Pa say, when he see dem carry his mammy off from dere, it make he heart swell in his breast."[18]

Another deplorable practice of slavery found in the Mars Bluff area was the breeding of slave women. Waiters told about the breeding practices that his grandfather recalled: "Pa say if you was a little man you didn't have no intercourse with no women. My granddaddy told me that. Didn't have no intercourse with no women. . . . That what he say now. Got to be a man then. [In answer to a question, Waiters said that the owners did not want to breed small slaves.] They take them [women] and carry them off to breed them. Just like you got your women here. You carry your women over to the other farm to breed them. He'll take his women and bring them to your farm to breed them."[19]

Ryer Emmanuel, who lived in the Claussen community, within a few miles of Gregg's home, recalled in a 1937 interview that the owners determined to whom women should be bred, though there was no moving them from farm to farm for breeding. Mrs. Emmanuel said: "Slavery chillun had dey daddy somewhe' on de plantation. Cose dey had a daddy, but dey didn' have no daddy stayin in de house wid dem. White folks would make you take dat man whe' if you want him or no."[20]

Josephine Bacchus, of Marion County, explained in an interview of the same year what a breeding woman was: "You see, dey would have two or three women on the plantation dat was good breeders en dey would have chillun pretty regular fore freedom come here." Ball told what constituted a good breeding woman. He said that the trader who brought him south also brought two pregnant women, who were the first people to be sold when they arrived in South Carolina. The slave trader boasted to the buyer that they "were two of the best breeding-wenches in all Maryland—that one was twenty-two, and the other only nineteen—that the first was already the mother of seven children, and the other of four."[21]

Former slave Hector Godbold, in a 1937 interview in Marion, said: "Den my gran'mammy use to run way en catch rides long de road cause de peoples let em do dat den. Coase if dey catch her, dey didn' never do her no harm cause she was one of dem breed 'omans."[22]

Waiters described the conditions for pregnant women: "Pa say work up to two months until going to find the baby. Then they let you work like you want to until after the baby come. Then give you a month after the baby born. Then you go back to work."[23]

Speaking of child care at Mars Bluff, Gregg had summed it up in a few words when he said, "Older woman, what done worked down, do most tending the babies." Ryer Emmanuel, who had lived several miles from Gregg, had much more to say about that:

> Oh, dey had a old woman in de yard to de house to stay dere en mind all de plantation chillun till night come, while dey parents was workin. Dey would let de chillun go home wid dey mammy to spend de night en den she would have to march dem right back to de yard de next mornin. We didn' do nothin, but play bout de yard dere en eat what de woman feed us. Yes'um, dey would carry us dere when de women would be gwine to work. Be dere fore sunrise. Would give us three meals a day cause de old woman always give us supper fore us mammy come out de field dat evenin. Dem bigger ones, dey would give dem clabber en boil peas en collards sometimes. Would give de little babies boil pea soup en gruel en suck bottle. Yes, mam, de old woman had to mind all de yearlin chillun en de babies, too. Dat all her business was. I recollects her name, it been Lettie.[24]

On some farms, children were probably cared for at a house some distance from the owner's house, but Mrs. Emmanuel said that she was cared for in the yard of the owner's home. She told of two incidents that confirmed the location: "My white folks was proud of dey niggers. . . . When dey used to have company to de big house, Miss Ross would bring dem to de door to show dem us chillun. En my blessed, de yard would be black wid us chillun all string up dere next de door step lookin up in dey eyes. Old Missus would say, 'Ain' I got a pretty crop of little niggers comin on?' De lady, she look so please like."[25]

Among the children in Miss Ross's yard was one mulatto child. Mrs. Emmanuel told a sad story about her.

> All de other chillun was black skin wid dis here kinky hair en she was yellow skin wid right straight hair. My Lord, old Missus been mighty proud of her black chillun, but she sho been touches bout dat yellow one. I remember, all us chillun was

playing round bout de step one day whe' Miss Ross was settin en she ax dat yellow child, say, "Who your papa?" De child never know no better en she tell her right out exactly de one her mammy had tell her was her papa. Lord, Miss Ross, she say, "Well, get off my step. Get off en stay off dere cause you don' noways belong to me." De poor child, she cry en she cry so hard till her mammy never know what to do.[26]

Gregg told Waiters about how the slaves' clothes were made. "Henrietta Dixon [the wife of the chief plowman] used to make clothes for them. Boss would go buy the cloth and her would make the clothes. . . . Blacksmith used to make the needle. Make needle hole right through there. Be about three inches long, some of them. Some two inches. Then you flatten it and put thread in it and sew."[27]

Sylvia Cannon gave a description of the African American's clothing.

Didn' get much clothes to wear in dat day en time neither. Man never wear no breeches in de summer. Go in his shirt tail dat come down to de knees en a 'oman been glad enough to get one piece homespun frock what was made wid dey hand. Make petticoat out of old dress en patch en patch till couldn' tell which place weave. Always put wash out on Saturday night en dry it en put it back on Sunday. Den get oak leaves en make a hat what to wear to church.[28]

Illness seemed to have been rare at Mars Bluff according to Alex Gregg's memory. Waiters said:

Pa say he ain't never know but two people to be sick and they had the typhoid fever.

People a little bit sick, but not enough to stop too much a work. . . . Didn't had much a doctors back in them days. Had to go in the woods and get your herbs to take. . . . [The boss] tell the man who feeling bad, he says, "Go out there and get some gourd-gut tea [tea made from the interior part of a gourd]."

Give you a half a teaspoon of that. Shucks. You just take it and rub your stomach with it. Pa say, "While you can do that, you done mess up the step."[29]

Waiters recalled what his grandfather had said about funerals for slaves. "If you sick and you die like this morning, you shut up in the house. They go on to work, work all day. They feed you and you go out and bury. That what my Grandaddy tell me. They work all day and bury at night with flambeaus." Ryer Emmanuel had seen such funerals when she was a child during slavery.

Couldn' bury dem in de day cause dey wouldn' have time. When dey would be gwine to bury dem, I used to see de lights many a time en hear de people gwine along singin out yonder in dem woods just like dey was buryin buzzards. Us would set down en watch dem gwine along many a night wid dese great big torches of fire. Oh, dey would have fat lightwood torches. Dese here big hand splinters. Had to carry dem along to see how to walk en drive de wagon to haul de body. Yes, child, I been here long enough to see all dat in slavery time.[30]

Both Alex Gregg and Ryer Emmanuel attributed the night funerals to the slaves' having to work all day—from sunrise to sunset. Only after coming in from the fields and eating could they bury their dead. A recent study by the South Carolina State Museum attributed night funerals to an African tradition that "spiritual contentment required that the sun be allowed to set on the open grave before burial."[31] Whatever the reason, there is no doubt that the custom worked to the advantage of the slave owners, who wanted to see the slaves working throughout the day.

Whereas Gregg said that he had seen only two people beaten, Mrs. Emmanuel told of seeing many people beaten.

> Oh, dey would beat de colored people so worser till dey would run away en stay in de swamp to save dey hide. But Lord a mercy, it never do no good to run cause time dey been find you was gone, dey been set de nigger dog on you. Yes, mam, dey had some of dese high dogs dat dey call hounds en dey could sho find you out, too. . . . White people would let de dog gnaw you up en den dey would grease you en carry you home to de horse lot whe' dey would have a lash en a paddle to whip you wid. Oh, dey would have a swarm of black people up to de lot at sunrise on a morning to get whip. Would make dem drop dey body frock en would band dem down to a log en would put de licks to dem. Ma was whip twice en she say dat she stay to her place after dat.[32]

Charlie Grant, who had lived on a large farm at the north edge of Mars Bluff, was quoted earlier about children being beaten if they were caught with a book. He also told of adults being beaten in his 1937 interview.

> All de slaves dat was field hands, dey had to work mighty hard. De overseer, he pretty rough sometimes. He tell dem what time to get up en sound de horn for dat time. Had to go to work fore daybreak en if dey didn' be dere on time en work like dey ought to, de overseer sho whip dem. Tie de slaves clear de ground by dey thumbs wid nigger cord en make dem tiptoe en draw it tight as could be. Pull clothes off dem fore dey tie dem up. Dey didn' care nothin bout it. Let everybody look on at it. I know when de whip my mamma. Great God, in de mornin! . . . I remember dey whipped dem by de gin house.[33]

Ball quoted an African American near Columbia to explain how slaves were tied up for beating.

> Each of my thumbs was lashed closely to the end of a stick about three feet long, and a chair being placed beside the mill post, I was compelled to raise my hands and place the stick, to which my thumbs were bound, over the top of the post, which is about eighteen inches square; the chair was then taken from under me, and I was left hanging by my thumbs, with my face

> towards the post, and my feet about a foot from the ground. My two great toes were then tied together, and drawn down the post as far as my joints could be stretched; the cord was passed round the post two or three times and securely fastened.[34]

Gregg, in his conversations with Waiters, had never mentioned patrollers—European American men who took turns riding horseback around the community to apprehend slaves who were away from their owners without a pass. Grant reported close contact with patrollers at a party at his grandmother's house.

> I was setting dere in de corner on dey blow cane. Common reed make music en dance by it. Dat de only way niggers had to make music. Dance en blow cane dat night at grandmother's house (Wilson place). Dey was just pattin en dancin en gwine on. I was sitting up in de corner en look up en patrol was standin in de door en call patrol. When dey hear dat, dey know something gwine to do. Dey took Uncle Mac Gibson en whip him en den dey take one by one out en whip dem.[35]

Ryer Emmanuel also recalled patrollers: "You see the colored people couldn' never go nowhe' off de place widout dey would get a walkin ticket from dey Massa. Yes, mam, white folks would have dese patroller walkin round all bout de country to catch dem colored people dat never had no walkin paper to show dem. En if dey would catch any of dem widout dat paper, dey back would sho catch scissors de next mornin."[36]

Gregg told Waiters that Sunday was their day of rest: "Worked six days a week—laid to rest on Sunday except going back to pull fodder. It be time to pull that fodder; you be working in the woods; then sometime on a Sunday have to pull the leaves off the corn stalk and tie it up for the mules and cows to eat." Gregg had said that he never went to church in a church building when he was a slave. Other former slaves in the Mars Bluff area had attended church with European Americans at Hopewell

Fig. 5 Slave gallery at Hopewell Presbyterian Church, in Claussen. The slave gallery, which can be seen in the upper part of the photograph, extends the entire length of the sides and back of the church. Many Mars Bluff inhabitants were members of Hopewell Church, and African Americans from Mars Bluff would have used this gallery.

Photograph by Barney Mattenson. Used with permission.

Church in nearby Claussen. Washington Dozier recalled, "De colord peoples go dere to dat same chu'ch en set en de gallery."[37]

Sara Brown, who was eighty-five when she was interviewed in 1937, said:

> De colored peoples worship to de white folks church in slavery time. You know dat Hopewell Church . . . dat a slavery church. Dat whe' I go to church den wid my white folks. I had a lil chair wid a cowhide bottom dat I always take everywhe' I go wid me. If I went to church, dat chair go in de carriage wid me en den I take it in de church en set right by de side of my Miss. Dat how it was in slavery time. Oh, my Lord, dere a big slavery people graveyard dere to dat Hopewell Church.[38]

Louisa Gause, who lived in Brittons Neck, in lower Marion County, found that it was a mixed blessing to attend the white people's church.

In a 1937 interview, she said: "When de colored people would get converted in dem days, dey never been allowed to praise de Lord wid dey mouth. Had to pray in dey sleeve in dem days. De old man Pa Cudjo, he got right one day in de big house en he had to pray wid he head in de pot."[39]

The words of former slaves that have been repeated here are painful to read. Still, they need to be recorded and read. For many years slaves had no way of recording the facts about their lives. Only through the pieced-together words of Alex Gregg and other former slaves can African American voices at last tell their own story about life in slavery at Mars Bluff.

3
Alex Gregg, Freedman

ALEX GREGG HAD lived twenty years in slavery when one day, near the end of the Civil War, a group of northern soldiers came to the farm where he lived. Gregg watched from the shelter of the woods as the soldiers entered the pasture.

Archie Waiters repeated his grandfather's story of what happened that day.

> Pa was here when slavery. They was fighting war, in slavery time. He said he was on a railing fence in the woods. They had an old railing fence, had cows in there.
>
> The people what fighting the war—they come through fighting war. Come along and kill master's cow.
>
> "Great God! They kill master cow! Master some hurted."
>
> And just set down and cook and eat, and they got their belly full. They didn't bother them [the African Americans]. They give them meat to eat and tell them they is free. The people ate so much it made them sick all around. They weren't used to eating like that. Used to eating collards and beans and peas and stuff like that; just give them enough meat to season it.[1]

If Gregg felt any joy when Sherman's soldiers told him that he was free, he did not mention it when he told the story to his grandson. Gregg no doubt knew about Amy Spain, a Darlington slave who had been too open in expressing her joy when she saw Sherman's cavalry arrive in Darlington. She had exclaimed, "Bless the Lord, the Yankees have come!" For that crime she was lynched.[2]

Waiters told what his grandfather did when the war finally ended.

> When the war was over . . . they didn't know what to do. Don't know where Missy and Boss was. Gone. Pa say he ain't see them no more. . . . Said he didn't know what to do. . . . Just like I living here on your place; don't see you no more. Go to next man. Don't see him. Go to next. Just like I want to stay there. [I want to] see you coming back. You ain't never coming back. Got split up somehow, the buckras. . . . Ain't never seen them people no more.[3]

> Pa said, "You know us supposed to get ten acres and a mule. That what us promised. When they fight the war to free the Negroes, promised ten acres and a mule.
>
> "Twenty-five cents an acre for land in my lifetime. Where you get twenty-five cents to buy land with? It takes all that for you to live."[4]

> Pa and Sis and Uncle Ruska [Gregg's sister and brother] leave and went somewhere back up there next the Seaboard Railroad. Then his sister died, and Uncle Ruska and my granddaddy come down here to Mr. Gregg place [the J. Eli Gregg place at the center of Mars Bluff].[5]

By that time Gregg was married to Emma Pigiott, and they had several children. They moved into a house on Back Street. Back Street was actually just a sandy road through a cotton field, with eight houses lined up beside the road. There were no trees—those would get in the way of the cotton. A man plowing there could pick up his plow as he walked across the little yards that surrounded the houses and go right on plowing. But Emma and Alex Gregg's house was different from the others. It stood at the end of Back Street and at the end of the cotton field. Larger than the other houses, it was the only one with its own well and two trees in the yard.[6]

In choosing to live on Back Street, the Greggs had chosen the liveliest spot in the whole community. Most of Mars Bluff was swamps and sandy fields, but there within hollering distance of Back Street was the commercial center of Mars Bluff. It was the place Mars Bluff people went if they went anywhere.

A general store presided over the intersection where the railroad crossed the main public road. On Saturdays, people came from miles around. As they approached the store, they looked wistfully toward the cotton gin and dreamed of the cool fall days when the gin would be running and the depot would be surrounded by bales of cotton—and perhaps they would have money in their pockets. Most of the people arrived on foot and left with their burlap bags slung over their shoulders. Maybe the bags contained a little coffee.

Across the railroad from the store, men got down from oxcarts and carried their precious bags of corn toward the gristmill. Above the noise of the mill, the blacksmith could be heard pounding on his anvil. All told, Emma and Alex Gregg had found about as dynamic a place to live as could be found in Mars Bluff.

Emma Gregg was quite dynamic in her own right. Waiters recalled: "They tell me her could pick 300 pounds of cotton a day. . . . A little small patch of cotton, pick this out in less than a day. Her'd walk over that thing and leave the sun running and go on to the next field. . . . When she wasn't on the sickly side, you couldn't keep up with her, they tell me."[7]

Mrs. Gregg had twenty-three children: Jimmy, Linda, Cornelia, Henry, Tank, Hannah, Spencer, Plum, Sidney, Houston, Emma, Cutter, Archie, Tena, Smilie, Maxie, Raymond, Lunda, Annie, Mac, and three others whose names Waiters could not recall.[8] Raising twenty-three children and being such an efficient farm worker would have been accomplishments enough, but Mrs. Gregg also had a full-time job as cook for the landowner's family from the late 1860s to the time her youngest daughter was old enough to take the job.[9]

Alex Gregg loved farming. Of course he would have liked to have land of his own to farm, but he had to be content with what he could get. At first J. Eli Gregg had wanted Gregg to work as a wage laborer in his fields, but Gregg was such a good farmer and had so many sons that eventually the landowner let him keep three mules—Jazz, Jessie, and Mary—at a stable at his house and sharecrop fifty acres, for one-third of the crop. Through all the years that Gregg sharecropped, his sons still had to work half-time in J. Eli Gregg's fields.[10]

Every Saturday night their mother would go to the store to get their pay. Waiters said: "Then they didn't pay off until Saturday night now. . . . About eight o'clock Saturday night. . . . Just like you my mother, you go over there [the store] and get that pay. . . . All [would]

Fig. 6 Business district of Mars Bluff, mid-1920s. An unidentified man hauls freight from the depot toward Gregg and Son's store, only the back part of which can be seen. J. Wilds Wallace is on the bicycle. The building on the right is a warehouse for the store.

According to Atleene Pinkney, the small building at the left of the warehouse was built in 1922 for Isaac Pinkney to sell meat, fish, and ice cream every Saturday.

Photograph courtesy of Amelia M. Wallace.

work. . . . She'd [Mrs. Gregg] go over there and get the money. . . . She'd get it for all of them. Anybody that worked for him, she'd get it."[11] The years passed with the whole family working on the European American's land. European Americans did not want African Americans to own land, so Alex Gregg was never able to have a farm of his own.

Decades later, Emma and Alex Gregg were sharecropping the same land when one day, about 1910, two men jumped off a freight car at Mars Bluff. The men, Otis Waiters and Robert McElveen, had set out from Lancaster, a hundred miles west of Mars Bluff, to seek their fortunes. Waiters was one of twenty-one children in a family that owned a farm near Lancaster, and McElveen was his friend.[12]

Fig. 7 Walter Gregg beside the railroad, about 1910. The bales of cotton stacked in the background indicate that Gregg is near the gin and the depot, at the east end of the Mars Bluff commercial district. On the west end of the clearing, not pictured in this photograph, were the store and the blacksmith shop.

Photograph courtesy of Amelia M. Wallace.

They went into the general store and asked if there was any work they could do at Mars Bluff. They were given jobs as farm workers and directions to an empty house they could live in. McElveen moved on within a short time to seek his fortune elsewhere, but Waiters stayed.[13]

Waiters met one of Emma and Alex Gregg's daughters, Tena. They were married and began their life together in a hewn-timber house that stood in a field less than a mile from Back Street.[14] That was where Gregg went to ask Tena and Otis for a baby "to pacify Emma." She was in poor health and lonesome in the house while everyone else was off at work, so the Waiterses gave their three-year-old son Archie to Alex Gregg.

4
Gregg's Grandson, Archie Waiters

ARCHIE WAITERS TOLD stories of his own life, stories that spanned more than a half century. That was the kind of information about early African Americans at Mars Bluff that had been sought in vain. For that reason, Waiters' life is recorded here. It is not one man's life story but a view of what life was like for African Americans at Mars Bluff in the first half of the twentieth century.

As a small child, Waiters took seriously his responsibility for the care of his grandmother. While everyone else was out working, he stayed with her. Asked what he did, he said, "Handing Ma water and stuff like that, and sweeping the floor, and scouring when I got big enough—and sweeping the yard."[1]

At seventy-one years of age, Waiters had few memories of childhood play; but he did recall having a "lee little wagon rim and a piece of wire bent to fit over that rim. . . . Push it and run and try to catch it. . . . All the children had one of them. Be racing up and down the road. See which one can outrun the other one."[2]

Waiters had even fewer memories of school: "I liked going to school but I couldn't go because I had to work. Out of the whole three months the school run, I wouldn't make but a week or two weeks or sometime three weeks."[3] Before Waiters was tall enough to reach the plow handles, he was plowing—holding the plow by the crossbar—helping his "Pa" sharecrop.

Waiters was a conscientious child, but making sweetening water was a temptation he could not resist.

> I'd make sweetenen water. Just put syrup in water and stir it up—a cup in more than a quart of water and stir it up. Sometimes I'd have to hide from the old man.
>
> "Boy, what you doing?"
>
> "Drinking some syrup water."
>
> Sometimes make so much I couldn't drink it. I'd go back later and drink it.

Waiters' joy in sweetening water was reminiscent of Charles Ball's pleasure in molasses. Ball told of the generosity of the woman in whose home he was assigned to live upon his arrival from Maryland. The woman gave him some molasses. "I felt grateful to Dinah for this act of kindness," he said, "as I well knew that her children regarded molasses as the greatest of human luxuries and that she was depriving them of their highest enjoyment to afford me the means of making a gourd full of molasses and water."[4]

Almost every day Waiters walked to his parents' home to visit them and his brothers and sisters. When the sun started to set, he regularly said, "It's time to go home to Pa's house"; and he returned eagerly to the old man whom he loved so dearly.

On the way to his parents' home, Waiters passed the house of his grandfather's brother, Ruska Gregg. Waiters would find his uncle Ruska nearby, at a spot in the woods where he worked, doing woodcraft. His old ox, Ned, who dragged up the logs for him, would be grazing at the edge of the field. Gregg would be working with an antique tool—a froe, a glut, an adz. Waiters would stop and stand very quietly watching the old man; he especially liked to watch him split and bend oak to make a basket. Gregg would sometimes ask Waiters to bring him a certain glut, and Waiters would run to get it from a box of gluts of assorted sizes.[5]

Waiters looked forward to the fall, when everybody picked cotton in the fields. Telling of it as an old man, his voice still reflected the excitement: "Everybody picking—everybody, men, women, children in the field. Babies in the field. Nobody home. . . . My mother tell me, 'All you pick over 200 pound, I'm going to give you and carry you to town with me Christmas.' I ain't never been to town in my life. I'd jump my neck for that."[6]

Fig. 8 Ruska Gregg's glut, used to split oak to make baskets like the white-oak basket pictured here. According to Waiters, Gregg made baskets for harvesting corn that were so big it took two men to carry them. When Gregg was very old, he gave this glut to Waiters.

As he grew into manhood, Waiters was of average height and build. His most outstanding characteristic was his facial expression—serious and discerning.

Waiters told about his first job as a man.

> The first public job I had was hauling slabs at Pasco's sawmill. I was walking the road one day and Mr. Pasco stopped me. He said, "Boy, you want to work?"
>
> He didn't have anyone to haul slabs at the sawmill so they had gotten piled up. Slabs, sixteen feet long, some twenty feet . . . I'd load them on a two-wheeled cart. The mule was hitched to it way yonder. No lines on that mule. . . . You tell him "Gee," and he'd just look back at you and go his way. He goes and stops at the right place and waits for you.
>
> When I got that job, I was making twenty-five cents a day. . . . I move all them pile of slabs, he raise me up to thirty cent a day.

Waiters later commented: "I remember when I worked, I couldn't draw my money. They wouldn't give it to me. They'd give it to Tena. She'd go on Saturday and get my pay. She'd give me a nickel sometime, a dime sometime; then I'd give it all back to her."[7]

The sawmill moved to North Carolina and summer came, so Waiters went back to work on the farm, plowing "sun up to dark."[8]

> Get up at five, catch your mule, water him, hook him up, and go to the field.
>
> Plow sixteen head of mule. Brother Big turned rows. . . . He give each man three rows. You plow your three, next man three, right on. . . . And then the row turner start his, way down yonder—and the crowd got to keep up with him. . . .
>
> Plow until noon. Carry your mule to the lot. Walk home. Us lived way up by the white people's church. Wash up and eat. Peas, beans, sweet potatoes, corn bread. Go back at one-thirty. Water your mule and plow until dark, dusk dark, just can see good, eight or nine o'clock.

When asked how much he made working on the farm, Waiters said, "Thirty cents a day. The row turner, Mr. Alfred Robinson, turned rows—he made a nickel more."[9]

Waiters described how he got a job at the cotton gin.

> I took Coker his breakfast to the gin house. Old man Sweeny didn't show up. Mr. Summerford asked me to work; I start to working in the seed house. . . .
>
> Mr. Summerford claimed he couldn't keep the gin house full of cotton; moved me out the seed house to feeding the suck pipe. I say, "I'll get it." [The suck pipe sucked the cotton out of the wagons into the gin.]
>
> Time I got there under the suck pipe, it take my cap off. Never saw it again. Gin cut it up.
>
> Talking about suck pipe won't do nothing. I got there, I say, "No wonder, got every air hole open."
>
> I shut it up, and, man, couldn't hear nothing but lint seed cotton flying through it—jug jug jug da rug, jug jug jug jug da rug. Then it cut off. The gin done get full, full up with cotton.
>
> Work at the gin just as long as you could see. Would be nine and ten o'clock at night, especially when the moon was shining.

Asked about pay, Waiters laughed and answered, "How much? That's slam out the question. I's making about two dollars a week."[10]

When Waiters was twenty, he married Catherine King, who came from the McCormick place in the northwest corner of Mars Bluff. She was a light-skinned, quiet, easy-going person.[11] The young couple set up housekeeping in a hewn-timber house near the home of Waiters' parents. The house had originally been on a street similar to Back Street, but all the houses had been moved so that each had a site of its own. The Waiterses' house sat in a pretty spot under a big oak tree at the edge of the woods.

Waiters' mother gave them a bed; King's mother gave them two chairs and a wood stove that had three legs and a stack of bricks for the fourth. Great-grandmother Irene Charles gave them an old dresser made of gopherwood. A shelf on the wall held their dishes, and a large fireplace heated the house. They had to tote water from Waiters' mother's house until a pump was put down in their backyard.[12]

Mrs. Waiters lovingly recalled that house where ten of their twelve children were born: "Use paper [newspaper] to keep the wind out of the cracks. Take flour and boil it and put paste over the paper and stick it on the wall. Put fresh paper on top of the paper already there every year

Fig. 9 Catherine and Archie Waiters beside their first home. Built in 1836, the house is of hewn timber with full dovetailing. Later construction at the rear added a kitchen and one small room.

about November. Beautify the front room—if you had books [magazines], get some of the pretty pictures and put on the wall. Stick them on."[13]

Archie Waiters recalled the year their first child was born; it was the 1930s—the depths of the depression: "Wasn't nobody working then. Nobody wasn't farming. . . . No money nowhere. No work. Everybody walking about. . . . Shoot marbles. All us grown people shoot marbles. Walk highway—pick up cigarette butts. The one find the most had to share."[14]

Being unemployed was an unusual circumstance for Waiters. He was very industrious and consequently was sought after when help was needed anywhere—the gristmill, the blacksmith shop, the railroad. But now there was no work anywhere. Waiters recalled: "That was the year we were working on the WPA. Worked about a month and a half, but you only worked half time. The other crew come and take it when the other crew knock off. I couldn't count all them people shoveling out that canal."[15]

Fig. 10 Archie Waiters harvesting sweet potatoes, 1930s. In the foreground is Ed Pinkney; on the ground are Archie Waiters and his brother James Waiters, and on the wagon, Waiters' brother-in-law Ernest Egleton and brother Tom Waiters. The mules are probably Jack and Bob. The baskets were made by Ruska Gregg.

Photograph courtesy of Amelia M. Wallace.

Times got better, and Waiters went back to farming. He recalled his favorite work, sharecropping: "Two acres tobacco, ten acres cotton. Give me three acres corn, my corn. . . . Me, my daddy, Ned Gregg, Edward Smalls sharecropped. All cured together with a string between owners. Put in on Monday; kill out on Friday. I got thirty-five cents for tobacco. I put fertilizer to make it. They said, 'Man!' . . . The sharecropper get a third. The other man get two-third."[16]

Mrs. Waiters recalled the days when her children were small. "Every day the children would sweep the yard and rake under the house. That where they play." Her greatest happiness was that all of the children were healthy. And she had a dream for them: "I wanted all of them to finish school and if they would, I would have liked for some of them to be a preacher."[17]

The Waiterses were members of the Church Aid Society, a lodge

Fig. 11 Catherine Waiters with yard broom she made to show how brooms were formerly made for sweeping the yard.

that provided burial for members and paid for a doctor when anyone was sick—but most of the children never saw a doctor.[18] Their births had been attended by a midwife, Maum Florence or Aunt Viney.[19] The most serious mishap Mrs. Waiters recalled occurred one day when she left four of her children at home and went several miles away to pick cotton. When she came back in the evening, she found that eleven-year-old Otis had been bitten by a rattlesnake and the younger children had not gone to get help for him. She took him to the hospital right away, and he recovered.[20]

Mrs. Waiters told of home remedies: "Every year in September, my mother would have the fever, and my mother-in-law, Tena, [was] the one that told her about the snakeroot. Hun [Archie Waiters] he went in the woods and got some and put it in a big pint of whiskey."[21] Waiters added: "A tablespoon in the morning and a teaspoon full when you get ready to lay down. . . . It bitterer than quinine. . . . You better have something to eat when you take that. It make you eat, give you an appetite. Fever got to go. Three days and that fever gone."[22]

There was sassafras root for measles and "quincy-like root that the old people used when the baby had diarrhea." Mrs. Waiters was not sure that she was saying the name correctly, but she was very clear on where to find it: "It grows in the corner there where you turn to go to Mt. Zion Church."[23] Waiters recalled his grandfather's purgative remedy: "Amy root. You dig the root up and you boil it and you drink the tea off it. . . . But you can't take but a teaspoon of it though."[24]

Mrs. Waiters said that it was the people in old Mars Bluff that made it so special.

> All the ladies used to get together and come to your house and they'd go to the store and buy white homespun cloth and you get the colored . . . and sew the pieces together. . . .
>
> When they gin the cotton at the gin house, you know that fragment cotton. We'd go there and get the people to give you that. . . . Put the cotton on the porch and take an old brush broom and whip the cotton until it get fluffy. . . . That what you pad the quilt with.
>
> And the ladies come to your house and make the quilt. Then

Fig. 12 Catherine Waiters with black snakeroot. Snakeroot is not as plentiful as it once was, but Mrs. Waiters remembered the place in the woods where it was most apt to be found. She walked directly to the spot and pulled this root.

Fig. 13 Before snakeroot disappears altogether, it deserves a close-up photograph. The fate of this root was to be poked into a bottle with a little whiskey.

Mrs. Waiters was not the only person who knew the spot in the woods where this snakeroot grew. One day a local forester working near the site saw a shiny new Cadillac with New York license plates drive into the woods. A dignified, nicely dressed man got out of the car, harvested a supply of snakeroot, and drove away.

Fig. 14 Catherine Waiters with medicinal plant formerly used to treat babies with diarrhea.

we go to their house and pull out their quilt. Everybody had their own quilting frame.[25]

Mrs. Waiters recalled the day that she and Sara Smalls and Gertrude Egleton went huckleberry picking "all down there most to the canal." They then went to town to sell the berries and walked halfway home before someone driving by stopped and picked them up. She also recalled working with her friends: "In November and December the ladies

Fig. 15 Catherine Waiters with quilt she made. Mrs. Waiters points at a long, narrow strip in her quilt and leaves unanswered the question, Does the quilt show an African influence?

would go in the woods and rake piles of straw to fertilize the cotton fields. In March they would scatter it across the fields. In May, start chopping cotton. September, pick cotton."[26]

There was plenty of work to be done at home too: gardening, cooking, soap making, lard making, crackling making, broom making.[27] Mrs. Waiters washed clothes in the washpot and heated the iron in the fireplace and cleaned it in the sand. She told how she made soap: "Old scrap grease what you fry chicken and fish in, you save it until you get a gallon or two gallons. Then you get four boxes of lye and just about three gallons of water and put it in the wash pot. Then you just stir it and stir it and it will jell."[28]

Mrs. Waiters had given up many of the old ways of cooking. She never made potato pone, though it was one of her mother's recipes, and she never made ashcake. She recalled her great-grandmother:

Fig. 16 Old ways of housekeeping. Mrs. Waiters recalls the years she used a flatiron for ironing, a broomstraw broom for sweeping the house, and a brush broom for sweeping the yard. She had a stove, so she did not need to use a three-legged pot for cooking in the fireplace.

She cooked ash cake. Had big old wide chimneys. Clean out the chimney good and burn oak wood. And then they would make up the meal dough and put fat back in it, and put it there and put the ashes over it. . . .

She used to talk about when the shake came [the Charleston

Fig. 17 Catherine Waiters with broomstraw broom. Mrs. Waiters never bought a broom; she gathered broomstraw and wrapped a long strip of old cloth around it, binding it together. This broom was made by Mrs. Waiters' neighbor, Jannie Pinkney, in the mid-twentieth century.

Fig. 18 Catherine Waiters demonstrating iron cleaning. Mrs. Waiters ironed for her family of twelve while living here. How many times she must have come out of the door into this yard to clean the bottom of her iron on the hard-packed white sand in her clean-swept yard.

> earthquake of 1886]. They calls it the shake, but now we call it earthquake. She said one night when her last baby came, that was when the shake came and . . . shake all the dishes out the cupboard.[29]

The lodge to which the Waiterses belonged met once a month, and the whole family went to Mt. Zion Church every Sunday. When asked if there were a lot of people there, Waiters said:

> What you talking about; church be full every Sunday. . . . They'd have Sunday School. . . . Then they get up and start to preaching. . . . The longest time we'd have when they'd get to taking up collection and the people start to singing and don't know when to stop. Don't know when to stop singing. Sing to get happy. You stop to sing; he take the thing and carry it right on. Used to was some good old singing back in them days. They'd sing about:
> Blow Gabriel, Judgement. . . .
> Blow Gabriel—judgement.
> Blow so loud he got to wake up the dead.
> Blow Gabriel, far from Eden.
> (That to wake up God's children. Means good people, like want to go to heaven.) . . .
> Blow Gabriel. Blow your trumpet.
> Loud and easy.
> (That to wake up God's children. Then—)
> Blow Gabriel.
> Blow it loud to wake up the damnation.
> (That mean the people what going to hell.) That's all I know. Walter Sellers used to love to sing that thing.

Mrs. Waiters added, "But he had a different way though." She sang:

> Blow Gabriel, judgement.
> Blow Gabriel, judgement. Lord.
> Sinners keep a'running—judgement.
> Sinners keep a'running—judgement.[30]

Mrs. Waiters recalled revivals in the spring and fall, when a guest preacher would preach every night for a week. Asked if she wasn't tired at the end of that week, she replied: "Would be when you'd pick cotton and you go home and you cook your supper and then get ready and then go. . . . Seem like you wouldn't be so tired. . . . Get up the next morning and you go to work." She sang from a revival song:

Oh Lord, I want you to help me.
Lord, I want you to help me.
When I'm on my way
Lord, Lord, Lord, I want you to help me.
Lord, I want you to help me on my way.[31]

In the 1940s, the man Waiters was sharecropping with stopped farming, and Waiters went to work cutting timber. He talked about working beside the sloughs in the river swamps, where he killed as many as "twenty-seven water rattlers" in one day. "Them a sad snake. Come with his mouth wide open. You better get out the way or kill him. Every tree you get, you better look around before you go up there. Every time you cut a tree, you kill a snake. Snake at edge of the water; you start to cutting and they come out the water at you; quit in the middle of cutting and kill the snake."[32]

Waiters said that snakes were not the worst part of working for the timber man. "Pay was something. He ain't going to pay you what he promised you. Give you your pay. Ain't see your money in an envelope. He just hand you your envelope and drive on off. You open your envelope and you ain't got a third of your money in there sometimes. You ain't got half of it in there sometime."[33]

Waiters tried working at a sawmill lifting massive sills but found the pay situation equally bad there.

> Had to have eight men to handle them sills. Ain't no weak man going to mess around with lifting them sills up. . . . Load on a gondola. . . .
>
> He wouldn't half pay you either. You be looking for your money. You be setting there waiting on that man. . . . He pay me something. He ain't pay me what he promised. . . . So me and Lacy Fleming went on down the railroad track—went on there [to the creosoting plant]. Put us to work the same day we

Fig. 19 Archie Waiters with his grandfather's picture. Waiters' expression tells the story of his happy memories of his grandfather. This picture was given to Otis and Tena Gregg Waiters. It travels from New York to Florence to Mars Bluff, each Waiters sharing it freely with the others.

> went there. Paying twenty cents an hour when I first start. Went to thirty cents an hour one week after I got there.

Shortly after he began work at the plant, Waiters was promoted to operating the doors on one of the pressure creosoting chambers.[34] At last he had a job that paid what he was promised.

In the 1950s, Waiters was presented with an unusual opportunity. Land had not generally been available for African Americans at Mars Bluff. Now a few lots were offered for sale. Many of them were too low for building, but Waiters obtained a good lot on high land, and he built a house on it. At last he was a homeowner—on land of his own.[35]

The late 1950s were an eventful time in another way. A new school was built for African Americans in Mars Bluff; until then there had been only a two-room schoolhouse. Also, for the first time, the school system supplied school buses for African Americans at Mars Bluff. Waiters' children now had access to the high school six miles away in Florence.[36]

Twenty-one years after Waiters started work at the creosoting plant, he was retired owing to physical disability. He had set a record—twenty-one years without being late to work or absent even though, at first, he had to walk four miles to work and, in the last years, he was badly crippled by painful arthritis.[37]

After retirement, Waiters enjoyed life at home with his wife, whom he still cherished as "Oman"—his name for her through the years. They were surrounded by grandchildren, for four of their children had built homes on nearby lots. It was then that Waiters had time to sit down and reminisce—to recall a lifetime of memories—to recapture part of the lost story of African Americans at Mars Bluff.

Part Two: *"It's Right There"*

Introduction to Part Two

I BACKED INTO my interest in rice. In fact, I had absolutely no interest in it until I heard an offhand remark by a forester. He said, "I know they used to grow rice around here, but no one knows how they grew it."[1]

Both the forester and I were aware of the old irrigated rice fields on the Harwell farm at the north edge of Mars Bluff. He was not talking about those fields; he was suggesting the existence of other fields in Mars Bluff. Why had I never heard of them?

I was curious. Rice planters in the tidewater region of South Carolina had been among the wealthiest people in the United States. If Mars Bluff had a similar history, I wanted to know about it. So the question stayed in the back of my mind.

Over the years, whenever I chanced to visit with people who might know the answer, I would ask them if they knew how rice had been grown at Mars Bluff. I asked all sorts of European Americans—historians, agricultural experts, landowners. The answer was always the same: "If you want to know about rice, go to Georgetown. That's where it was grown."

It was hardly surprising that no one knew anything about how rice had been grown at Mars Bluff. It was not a likely-looking place to raise rice. It was a sandy cotton-farming community forty-five miles above the Georgetown tidewater region, where the large, irrigated rice plantations had been.

I despaired of ever learning anything about rice at Mars Bluff—but I was underestimating Archie Waiters.

5
The Discovery

ARCHIE WAITERS WAS sitting in a low chair on his porch, holding my tape recorder in his gnarled hands. Below us in the yard, a red hen clucked and scratched in the hot sun. A little boy walked over timidly from a neighboring house to sit on the steps and listen to his grandfather. Waiters was explaining how the old-timers split a rail.

"You take a wedge and a maul," he said.

I interrupted him. "I've never seen a maul. Could you make one?"

"Sure I can make a maul. You make it the same way you make a rice maul."

When Waiters said the word *rice,* I could not believe my ears. I leaned forward so that I could hear him better. Did he know something that no one else knew?

"What do you know about rice mauls?" I asked.

"I've made a rice maul. I used to help Willie Scott grow rice. I can show you the field."

I could not believe it. After years of fruitless inquiry, I had stumbled upon a man who could tell me how rice had been grown. And he could show me the field.

We set out in the car to go see Willie Scott's rice field, heading south. I thought that Waiters would tell me to turn at the road that led down into the swamp, but we went past it.[1] We were still crossing sandy fields, within sight of the landscaped grounds of a local college, when Waiters said, "Turn here."

Surely we were not going to find a rice field here. We crossed a sandy field and entered the cool shade of the woods. Waiters leaned toward the car window and pointed to an old ditch bank that was barely visible through the trees.

"There's the ditch," he said. "Stop here."

When the car stopped, Waiters grasped the door frame and pulled himself to a standing position. His face was ageless. Standing beside the car, thin and erect, he looked almost like the young man he had been a half century earlier when he was one of the best plowmen at Mars Bluff.

He moved through the pine-scented forest with a quiet dignity. But his steps were quicker than usual. He was hurrying to prove that he had spoken the truth, for he knew that I was having trouble believing that he knew how rice had been grown.

Waiters came to the shallow ditch. Finding a place where the banks sloped gently, he climbed carefully into the ditch and out the other side. I followed him. Standing amid the tall gum trees that now covered the area, he said, "This is the rice field." We stood on a low embankment by the ditch and looked at the sixty-year-old forest that had grown on the spot that had once been a rice field. The land was low here, and water collected naturally.

"We didn't have to water the rice if there was enough rain," Waiters said. "If the water got a couple of inches below the top of the bed, then we'd water it. We had a pulley over the well, and two buckets. One man would hoist the water and the other man would pour it on the field."

Waiters searched along the ditch until he found a place that was a little wider and deeper than usual. "That's the well."

Looking at the shallow depression in the ditch, Waiters seemed to relive a day when the well was full of water. "I draws a while. He catches and pours out. I get tired of pulling up. He pulls while I pour. We stop watering when the water runs over the top of the beds." Waiters laughed. "And just about time we'd get done and start back to the house, we'd hear the thunder, 'RARARARA,' and there'd come a big rain."

Waiters and I followed the little embankment that ran beside the field and immediately arrived at the opposite side. The field was so tiny, about 120 feet square—less than a third of an acre.

Cutting across the field to its lowest side, Waiters pointed to three cuts in the embankment that Willie Scott had made one fall day, sixty years earlier, when he had drained the water off his last crop.

"After we'd drain the field, the rice would start to lay down." Waiters spoke as if it were just yesterday that Scott had drained the field; I could almost see the golden rice around us.

"We'd take a reef hook," he said, "and cut it and hang it on a drying

Fig. 20 Drawing water for the rice field. Standing in the biggest ditch at Willie Scott's rice field, Waiters shows how they used a pulley to hoist water when there was insufficient rain to flood the field.

Photograph © 1987 Sidney Glass.

Fig. 21 Pouring water onto the rice field. Still standing in the ditch, Waiters shows how they poured water into a trough to carry it over the embankment and into the field.

Photograph © 1987 Sidney Glass.

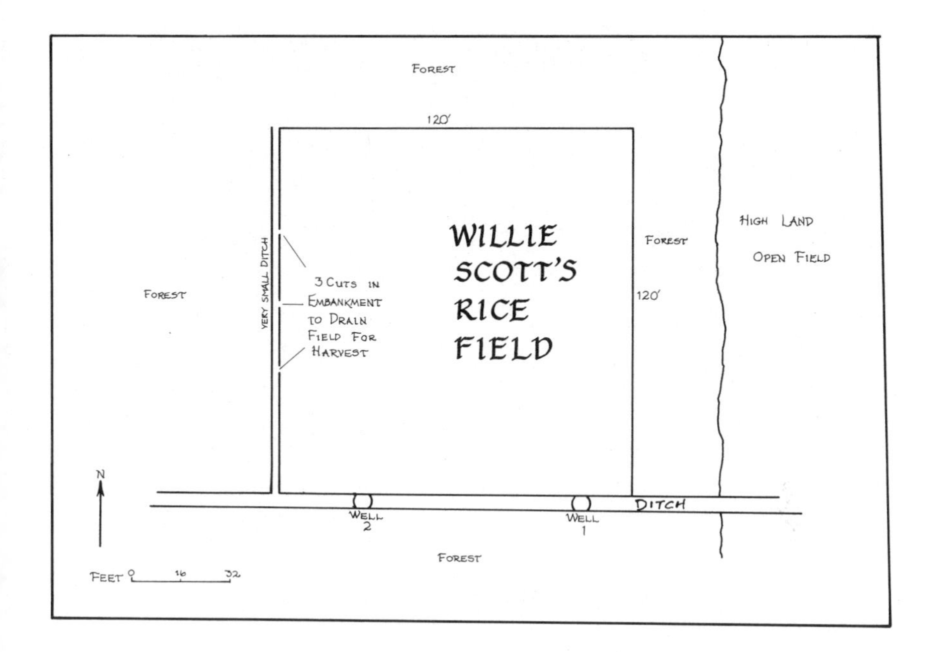

Map 5 Willie Scott's rice field. Well 1 was used when rain was not sufficient to keep the field flooded. Well 2 was available for use but was never needed. An embankment of approximately one foot surrounded the entire field.

Map by Dinah Bervin Kerksieck.

rack in Willie Scott's yard. Some of the rice would drop off, and we'd beat the rest off with sticks. Then you put it in a sack and give it away."

Give it away? Where did that idea come from? Certainly not from the thrifty Scotch settlers of Mars Bluff.[2]

After we had left the field, Waiters said a strange thing. I asked, "Who did you do that with?" I meant who was working with him when he was flailing the rice, but he interpreted the question to mean what European American landowner was involved. He replied, "That used to be Willie Scott's down there. He don't need no white folks."[3] Waiters was telling me that though every other crop had to be sharecropped with the landowner, rice was different. The African Americans raised it on their own.

Willie Scott's field was just the beginning. Waiters knew of other African Americans who had grown rice at Mars Bluff.

Frank Fleming, a sharecropper with seventeen children, had been a highly respected resident of Mars Bluff for the better part of his life.

Fig. 22 Draining the rice field. Standing beside a tiny ditch on the lowest side of Scott's rice field, Archie Waiters shows how they cut small trenches through the embankment to drain the field. Beyond Waiters' shovel is a cut in the embankment that they made sixty years earlier when Scott harvested his last crop of rice.

Photograph © 1987 Sidney Glass.

Fig. 23 Waiters standing on the former site of Alex Gregg's rice field. The present farmer has tried to drain the area to make it part of a corn field, but the large gum trees show that it is better suited for rice than for corn.

Waiters said that Fleming had raised rice until he moved away from Mars Bluff about 1920.

To get to Fleming's rice field, Waiters had me turn at a road that sloped gently down into Polk Swamp. But we did not go to the bottom of the swamp. As soon as the land dropped off to a wet bog, Waiters pointed to a forest of gum trees and said, "Stop. That's the field."

We walked into the woods on the high side of the old field. The air

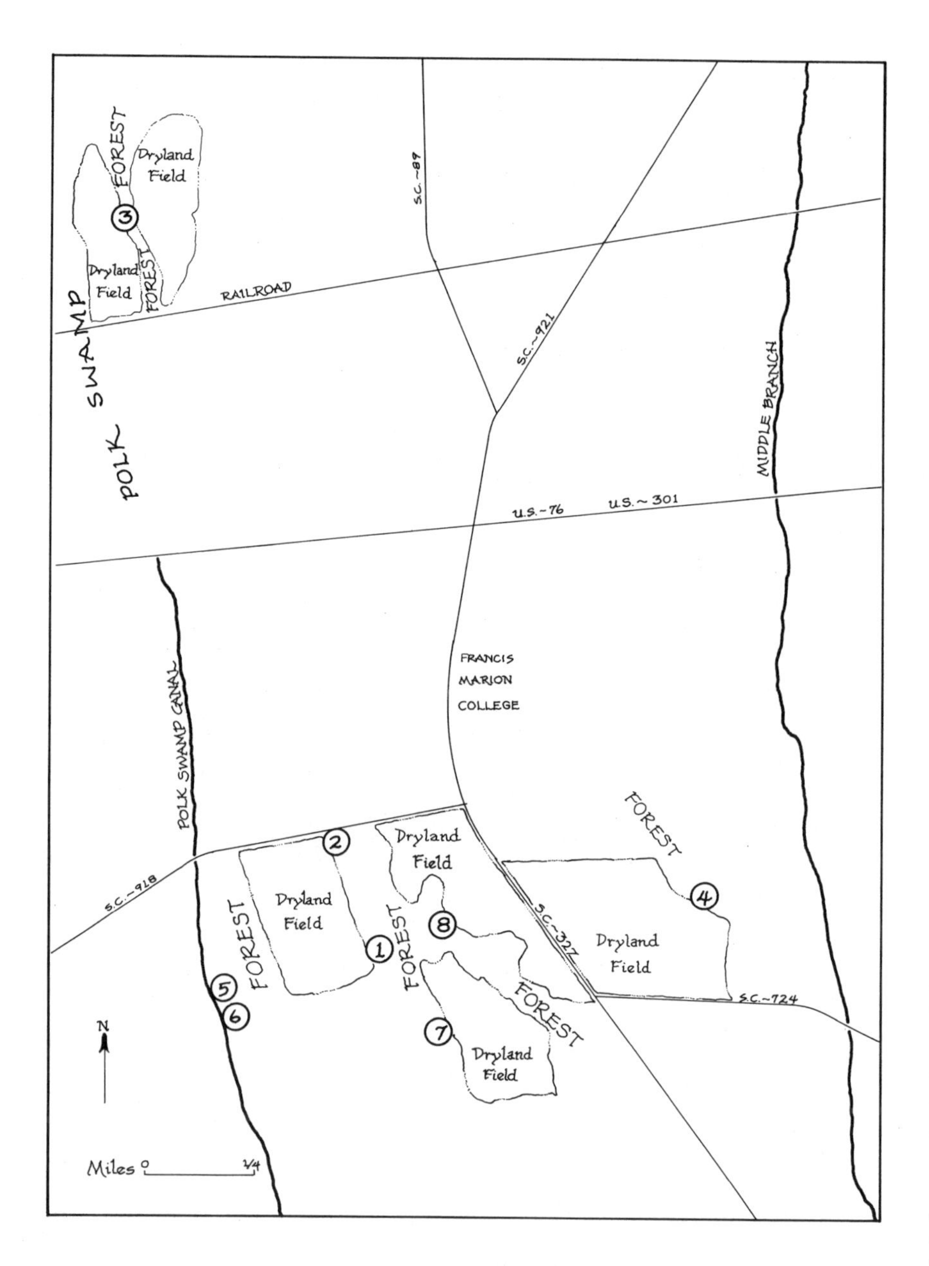

Map 6 Eight rice fields known by Waiters. The rice growers tended to choose land immediately adjacent to high land. Six of the eight rice fields follow that pattern. Only the Hayes brothers chose land that was surrounded by swamp.

Map by Dinah Bervin Kerksieck.

1.	Tom Brown	5.	Jim (?) Hayes
2.	Frank Fleming	6.	Willie (?) Hayes
3.	Alex Gregg	7.	Jim (last name unknown)
4.	Alex Harrell	8.	Willie Scott

Map 6 Key

was heavy and smelled of moist earth. To our left, the dry land sloped gently up to a forest of beech and oak; to our right, the old rice field, now overgrown with gum trees, lay low and wet.

The field was almost as wide as Willie Scott's and twice as long. Waiters said that enough rainwater collected here for Fleming never to have to water his rice.[4]

Beyond Fleming's rice field was a strip of high land where he had raised corn. And beyond that, where the land became swampy again, a very old man named Tom Brown had grown rice until he died in the 1920s.

Besides these two rice growers, Waiters told of four other men, as old as Brown or older, who had raised rice nearby. All had small fields, less than an acre, and all depended entirely on rainfall to keep them flooded. Like Fleming and Brown, Jim (last name unknown) had located his field where the land first drops off into Polk Swamp. Two brothers named Willie and Jim Hayes had had fields immediately adjacent to Polk Swamp Canal. Still, according to Waiters, they relied on rainwater, not the canal, to flood their fields. These three men grew rice until their deaths. Their fields were abandoned by 1920. The fourth man, Alex Harrell, had located his rice field near his home, on lowland at the outer edge of the swampy area surrounding Middle Branch. He raised rice until he moved away, also about 1920.[5]

When we got in the car to go see Alex Gregg's rice field, Waiters, who generally wore a serious expression, smiled in a peaceful way, recalling the love that he and his grandfather had shared. We went about a mile north of the other rice fields and passed a big oak near which Gregg's house used to stand; then Waiters told me to drive down a steep hill into Polk Swamp. There, beside a corn field, we came to a spot where tall reeds grew, and Waiters said that this had been his grandfather's rice field.

Why had Gregg chosen this spot? When I got home, I studied a soil map. Gregg had chosen for his rice field the only perennial pond in the area. No wonder Waiters told stories of his grandfather having four mortars and pestles—the more water, the more rice.

Gregg had stopped raising rice by 1920. His twenty-three children had grown up and found work. There was no one left to help him, and he was too old to cultivate rice by himself.[6]

Altogether, Waiters had identified eight African Americans who had cultivated rice until about 1920. I had come a long way from believing what the European Americans had told me, that no one knew how they used to grow rice at Mars Bluff—and I would never have guessed that this was only the beginning of the surprises I would stumble upon.

6
Surprises

HAVING DISCOVERED THAT Archie Waiters knew about rice cultivation, I was curious to know if other African Americans at Mars Bluff knew people who had raised rice. I wanted to pursue that story, but I had no time to interview people about rice. Each year I had only a few weeks in South Carolina, and I had promised myself that I would devote my time to recording the story of the Rosenwald school in Mars Bluff. That story would soon be lost when the older generation of African Americans died, and I was the only person who had the time and the interest to rescue the facts before they were gone forever.

The Rosenwald school in Mars Bluff, built in 1924, was part of an exciting two decades of school building in the rural South. The building program started in 1913 when Julius Rosenwald, of Sears, Roebuck and Company, agreed to help build six rural schoolhouses for African Americans in Alabama. The six buildings were such a success that Rosenwald agreed to help build a few more—and then a few more—and then more—and more. By 1932, he had contributed to the building of 5,000 schools.

Rosenwald always stipulated that local African Americans and state school systems match his donations. I had read stories about how African Americans in Alabama had raised money for their schools. One former slave said he wanted "to see the children of my grandchildren have a chance," so he gave his life's savings, a little greasy sack containing pennies, nickels, and dimes. "One old woman in Alabama roused a meeting by giving her only money, one copper cent."[1] If Mars Bluff African Americans had similar stories, I wanted to record them. Rice would have to wait. So I used my limited days at Mars Bluff interviewing people about the Rosenwald school.

Those interviews produced strange results. The people who had built the school had died, but I thought that their children would recall how their parents had raised the money. Seven interviews led me to the conclusion that in the 1920s African American adults at Mars Bluff did not share their adult concerns with their children. The children, now in their sixties, knew virtually nothing about how money had been raised to build their school; few had heard their parents discuss the matter.[2]

Why had they not heard it discussed? The answer to that question may lie in a comment made by one of the people interviewed. He was talking about proper etiquette for children in that day: "If grown people were talking and you walked up and then be looking at their face talking" and "looking in their mouth, taking in what they say—if they had a conversation, you better stay your distance."[3] In other words, parents did not allow their children to hear them discuss business, and wondering if they would succeed in getting their schoolhouse was very serious business. Consequently, my interviews produced little about the building of the Rosenwald school.

Fortunately, I had added one question to the school-building interview. After asking questions about the Rosenwald school, I would say, "Did you ever know anyone who raised rice at Mars Bluff?" It was an inefficient way to conduct a rice survey, so I was amazed at the replies I received.

Claudia Williamson Williams (1911–1985)

Claudia Williams was a retired teacher who had taught at the Rosenwald school. Her home was in Florence, next door to a garden club park that she had promoted. She was a gentle, soft-spoken person—the kind of teacher that students love. When I arrived, she was worrying, afraid that she might not know the answers to all my questions.

Mrs. Williams was born at Mars Bluff but had left while still quite young to board in Orangeburg, the site of two African American colleges, where she could get a good education. Her entire life had been spent in education; so she was not a likely person to know anything

Fig. 24 Claudia Williams, standing in front of the Rosenwald school where she had taught.

about agriculture. But when I asked about rice, she said, "They used to raise rice over there where my grandfather lived."[4]

Her grandfather, Tony Howard, had had the distinction of being one of the African Americans who served in the South Carolina legislature after the Civil War.[5] He had owned a fairly large farm, and Mrs. Williams recalled that he had had two mortars and pestles—an indication that he had raised a fair amount of rice.[6]

I wanted to see Howard's rice field, so Mrs. Williams drove with me to Mars Bluff, though she had misgivings about being able to find the old road to her grandfather's place. When we became frustrated looking for the road, we stopped at a house. A bright-faced young woman, Lena Mae Douglas, came out, and Mrs. Williams asked, "Where is the old road?"

Smiling broadly, Mrs. Douglas put her arm around Mrs. Williams' shoulders and pointed just beyond our car. She said, "It's right there."

Like so much of the African American story, it was right there

Fig. 25 "It's right there." Her arm around Mrs. Claudia Williams' shoulders, Mrs. Lena Mae Douglas points to the old road leading to Tony Howard's place.

and we had not seen it. Even with Mrs. Douglas' good-humored help, we never did find the old rice field.

Matthew Williamson (1915—)

Mrs. Williams asked her brother Matthew Williamson, an imposing-looking retired school principal, to lead me to Tony Howard's place.

All that was left of Howard's home site was one tree standing alone in a field. Nearby in the woods, Williamson pointed to lowland beside a ditch—the old rice field. Today the area is covered by large hardwood trees, but Williamson described how the rice field used to lie beside the ditch, about three hundred feet long and sixty feet wide. He recalled seeing it when he was a child:

They had several places dug out along the ditch by the rice field. . . . You could go down there and take your bucket with a rope on it and throw it in the ditch and then pull it back to you and take the water and pour it in the field. . . .

I think my uncle, Robert Gregg, planted a few years after he lived here after my grandfather passed. . . . Must have been 1920 or 1922 was the last time they plant the rice. . . .

. . . my father [Sherman Williamson] planted a little rice up there on his farm. . . . He had one or two rows. I remember vividly working in it. We had a spring there in a ditch, and we would pull water from that ditch and pour it in the field . . . when it got dry. . . .

. . . that must have been about 1922. . . . At that time, we had seasons when it was so wet that you didn't have to worry about pouring or dipping water from the spring.[7]

Williamson told me about his father, who was a very active man. Sherman Williamson farmed his own land, was a carpenter, did butchering, and drove his wagon (in later years, his car) through the country selling meat. He served as trustee of the school, as secretary of Mt. Zion Church, and as choir leader. His wife, Mary Howard Williamson, was a daughter of Tony Howard.[8]

MABLE SMALLS SELLERS (1908–1990)

I found Mable Smalls Sellers at her son's home in Florence. A large woman, she was confined to the house because of poor health, but her mind was still very sharp. When I asked if she knew anyone who had grown rice, she replied: "My daddy used to grow it. He . . . planted rice about 1922 . . . on the side of the forty-acre field [a swamp corn field]. . . . I guess about a half acre. [It would be wet] most of the time. That's why he planted it on that side. Be wet places."

Mrs. Sellers said that her father, Lewis Smalls, used a reef hook to cut the rice and a mortar and pestle to husk it. To remove the grains

Fig. 26 Sherman Williamson and Mary Howard Williamson.
Photograph of unknown date is courtesy of Matthew Williamson.

Fig. 27 Former site of Sherman Williamson's rows of rice. Matthew Williamson stands on a bridge over the spring from which, about 1922, when there was not enough rain he dipped water to pour on the "few rows of rice" planted beside the ditch. The rice was conveniently located near Williamson's home, a corner of which can be seen in the background.

from the stalk, he would "take a few strands and . . . strip the rice off [with his hand]."

Smalls raised rice for only two or three years around 1922. When asked why he quit, Mrs. Sellers said, "I don't know. I guess he didn't figure it paid off."[9]

Mattie Smalls Gregg (1910–1989)

Mattie Smalls Gregg, one of Mars Bluff's best retired cooks, had answered all my questions about the Rosenwald school, and it was time for her to start cooking. Her son was coming over to eat dinner with her. I should have left, but I said, "Do you remember anybody planting rice?"

"Oh yes," she said. "Oh yes, Alfred Robinson."

Fig. 28 Mattie Smalls Gregg, the second adult from the left. According to Waiters, the county school system did not provide Mars Bluff African American children with school-bus service to the nearest high school in Florence, so the parents raised money and bought this bus. They also paid for gas, repairs, and the driver's salary. Apparently the Mars Bluff people were making use of their bus for an outing to Florence when this picture was taken about 1949.

Photograph courtesy of Amelia M. Wallace.

Mrs. Gregg had seen Robinson's rice field in the 1920s. It was south of Tom Brown's rice field and in the same sort of location, at the edge of the swamp where the dryland first drops off into the wet lowland.

Mrs. Gregg laughed, recalling a long-forgotten joke: "My daddy used to plant . . . a little corn field, back up in there. And Alfred . . . used to plant that rice field. . . . Alfred be down there plowing and cutting the old mule and making the old mule go."[10] Mrs. Gregg laughed heartily, recalling what Robinson would say to the mule—but she never would tell me.

FRANCES MISSOURI JOHNSON (1915–1991)

Frances Johnson had the distinction of being the child who had walked farthest to attend the Rosenwald school—four miles to school and four miles home.

At age seventy, she still lived with her sister and brother on her father's old farm. It was in the neck section of Mars Bluff, the V of land made by the junction of the Great Pee Dee River and Jefferies Creek.

I found Miss Johnson at her home, a new white house beside an ancient oak. She was young looking and quick moving, and her eyes twinkled behind horn-rimmed glasses. Every now and then while we talked, she responded to my question with a peal of laughter and an incredulous, "Now, what do you want to know that for?"

When I asked, "Do you remember anyone planting rice in the neck section of Mars Bluff?" she said, "My father used to plant rice."[11] And immediately she was telling about strange practices—beating the rice on a scaffolding and making a "frail bar."

Describing the flail that her father, Vico Johnson, had made, she said, "You burn that thing and you twist it. Twist it like that, so it would flop kind of like a chicken head when you wring it."[12] What was she talking about? A flail is a wooden implement used to beat grain off the stalk. Archie Waiters used any stick. Some Georgetown plantations used two pieces of wood tied together with a leather thong. No one had ever heard of the flail that Miss Johnson described.

Although she had never made a flail herself, she agreed to show how her father had made one. She took a small hickory sapling and put the center of it in hot coals. When she took it out of the coals and pressed the tip of it on the ground, she was disappointed that it did not become flexible at the center as her father's flail had. Still, she had shown how her father had used hot coals to make a flail—a technique that, so far as I have been able to learn, is known to no one else.[13]

Miss Johnson described another of her father's unique practices. Vico Johnson put six short posts in the ground, and on these he built a tablelike rectangular scaffold made of saplings spaced six inches apart. He laid the rice on the scaffold for flailing; the grains of rice fell through to a sheet spread on the ground. It was an unheard-of procedure—no doubt meant to accomplish some winnowing during the threshing of the rice.[14]

Miss Johnson told me of another unusual procedure. When she had husked rice for her mother, she had put corn shucks in the mortar with the rice—to make it white.[15] What a day of surprises this was: three customs that I had not known existed.

What is more, Miss Johnson gave a clue that led to an unsolvable mystery. Her sister was named Julia Johnson, and her brother Chalmers Johnson. Those were prominent names in Dr. William Johnson's family. But Frances Johnson said that she did not know anything about Dr. Johnson or the irrigated rice fields on his farm, or where in Mars Bluff her father had been raised. So the question remains unanswered—had Frances Johnson's family once worked in those irrigated rice fields?

Miss Johnson showed me the field where her father had planted rice until about 1939. It was the usual size, 120′ x 150′, on low, moderately wet land. It was located beyond her backyard, past a pump and a small barn. The barn contained an item I had never seen in Mars Bluff before—the reef hook that her father had used to harvest rice. Johnson had gotten it from his father-in-law, Lewis Charles, who had used it for harvesting rice. Charles was a Mars Bluff man, so it seems reasonable to count him among the Mars Bluff rice growers. Miss Johnson, however, never saw his rice field.[16]

Leon H. Coker (1913—)

Leon Coker lived in a new house on the farm that his father had owned near the center of Mars Bluff. His father had bought the piece of land

Fig. 29 Frances Johnson with rice flail she made. She says that this one is too large in diameter and does not flop like her father's did.

Photograph © 1987 Sidney Glass.

Fig. 30 Frances Johnson flailing on a scaffold. This poor photograph may be the only one that will ever be obtained that demonstrates Vico Johnson's flailing technique. The scaffold is deficient in two respects: the corner posts are a foot too short; and the crosspieces are too few in number and therefore placed too far apart.

Miss Johnson shows the correct position for flailing. No one else I interviewed at Mars Bluff had heard of a tablelike scaffold for flailing rice, and all assumed that one stood at the end to flail. However, if the rice sheaves are pictured lying across the crosspieces, then Miss Johnson's position at the side of the scaffold is seen as logical.

Fig. 31 Frances Johnson with a sickle used for cutting rice, known in Mars Bluff as a reef hook.

from a lawyer who had received it in payment for settling an estate in the 1920s.

The senior Coker had been active in school affairs, and Leon Coker had been a sharp-eared little boy. He had even heard conversations between his parents when his father came home from the meetings that planned the building of the Rosenwald schoolhouse.

Fig. 32 Leon Coker. His father planted rice at Mars Bluff in the 1930s.

When I asked Coker my question about raising rice, he replied: "Well sure. We raised it ourselves. . . . my father, Preston Coker. . . . Must be around '33, '34, and maybe '35. . . . My father always would say anything would come along, he would try to make something out of—if he could. He just say he was going to plant that bottom in rice. . . . This bottom field would hold water during this time of year, depending on how much rain you get."[17]

Why had Coker started planting rice as late as the 1930s? Was it a deliberate continuation of an African American tradition, or was it a response to the United States Department of Agriculture's emphasis on farm self-sufficiency during the depression?[18] No one knows. Whatever his motivation, Preston Coker was apparently the last African American to begin rice cultivation at Mars Bluff. Vico Johnson, however, who had planted rice long before Coker started, continued cultivating it for years after Coker stopped.

By the time I finished the interviews—about the Rosenwald school—I had come to believe that rice cultivation was a well-established practice among African Americans at Mars Bluff. Out of seven people interviewed, six had known someone who had raised rice.

As remarkable as those numbers were, I had found the individual people that I had come to know even more interesting than the numbers; and now I was looking forward to becoming better acquainted with one of my favorite characters from Mars Bluff's past, Fanny Jolly Ellison. I only knew her through Archie Waiters' stories, but she had always seemed such a vital, lively person. I was looking forward to a visit with her daughter, Ida Zanders.

7
A Strange Place for Rice

IDA ZANDERS LIVED in Florence in a little green house behind a bright red crape myrtle. She answered the door and walked slowly back to her chair, explaining that she was slowed down by heart trouble. But she spoke in an energetic, laughing voice, so apparently her spirit was still strong.

I had come to learn about her family because they had lived in one of the historic hewn-timber houses at Mars Bluff. These were the same houses that Archie Waiters and I had discussed. I was still trying to gather information to support my claim that the houses should be saved.

Mrs. Zanders told me about her mother, Fanny Jolly Ellison, who proved to be just as lively a soul as she had always seemed in Archie Waiters' stories. And Mrs. Zanders talked about her twelve brothers and sisters and her father, Hillard Ellison, who had done farm work at Mars Bluff before moving to Florence to work in the railroad shop.

Although Mrs. Zanders had lived in Florence for most of her seventy-five years, she had great tales of Mars Bluff, complete with ghosts and children screaming like panthers. She was telling about the old people who had been the Ellisons' neighbors at Mars Bluff, and I asked her if she remembered any of those old-timers planting rice.

> Unhu. My momma plant rice right there on Palmetto Street [a main thoroughfare in Florence].
>
> *Q.* You're kidding.
>
> *A.* Fanny plant rice. Made me and my husband—the same man, I was courting him then—used to have to go out there and get it and beat it. . . .

Fig. 33 Ida Zanders standing on the spot where her mother grew rice when she lived at Mars Bluff. The site is now at the center of the Francis Marion University campus.

> Whilst I was courting. If he come by there in the evening, "Ollie, you and Buck go out there and get that rice." Well, I had to do it. . . .
> Q. Would she have her vegetable garden there [on Palmetto Street]?
> *A.* Right there.
> Q. Would she plant the rice right next to the vegetables?
> *A.* Unhu. I believe she plant the rice first. But I know I's beaten major rice right there. . . . I stop him from coming to see me [laughing]. . . .
> Q. Oh. . . . She'd just make you [beat the rice] if he was there?
> *A.* If he was there, because he was a boy. . . .
> Q. Did your mother have rice out at Mars Bluff?
> *A.* . . . Oh, yeah. Because my momma didn't buy nothing.[1]

Mrs. Zanders went with me to Mars Bluff to show me where her mother's rice patch had been. At the center of the Francis Marion

Fig. 34 Fanny Jolly Ellison (1878–1943), the only woman among the known rice growers at Mars Bluff.
Photograph, after 1926, courtesy of Ida Zanders.

University campus, in front of the library, we stood on the spot where her mother's house had stood. Mrs. Zanders said: "Go out the back door. . . . The rice was over here. And the beans and the peas, the collards and the peppers. . . . It was black dirt. . . . Get all the okras you want."[2]

Apparently the rice was planted on the lowest side of the garden in moist black soil. However, the spot in Florence where Mrs. Ellison grew rice was high land where no water would have collected. So Mrs. Ellison held the record for having the driest dryland rice field of any I saw.

Mrs. Ellison held another record. She was the only female rice grower I was to find among the numerous rice growers at Mars Bluff. That might be significant. Why in this family was Mrs. Ellison and not Mr. Ellison the rice grower?

In some African ethnic groups, male rice growers were the norm; in others, rice was the woman's crop, and men would have nothing to do with it. Judith Carney wrote: "Among Gambian ethnic groups, rice has been traditionally grown by women."[3] It is possible that the Ellison family came from a tradition differing from that of the other Mars Bluff rice growers—one that held rice cultivation to be women's work.

Mrs. Ellison may have a third distinction. Although I have made no study of rice growers in Florence, I would speculate that Mrs. Ellison was one of the last people to plant rice within its city limits. She had moved from Mars Bluff to Florence in the 1920s, and she planted rice on one of Florence's main thoroughfares well into the 1930s.[4] Those dates suggest that Mrs. Ellison was planting rice after most people in the area had stopped. So she may have been the last known rice grower in Florence.

8
HESTER WAITERS

HESTER BAILEY WAITERS (no relation to Archie Waiters) was my only hope for solving the mystery of the old rice fields on the Broadwell farm in the southwest section of Mars Bluff.

Joseph Broadwell, the owner of the farm, had shown me where three abandoned fields were located just inside the woods, at the outer edge of Jefferies Creek Swamp. Two of the fields were tiny and were situated at the outer edge of the swamp, where the high land first dropped off into lowland. The third field was much larger, 318′ × 270′, and was located a little farther into the swamp. Broadwell recalled that when he was a boy old African Americans always called these "the old rice fields."[1]

Had these fields belonged to African Americans or to European Americans? Broadwell insisted that everyone who knew the answer to that question was dead, but he directed me to the oldest living person from the Broadwell farm, an African American, Hester Waiters.

I found Mrs. Waiters in a wheelchair at her home on a quiet street in Florence. She was a thin, bright-faced woman with a most expressive voice.

Mrs. Waiters had moved to the Broadwell farm only after she had married, so her memories did not go back as far as I had hoped. Also, she knew nothing about the abandoned fields in the woods on the east end of the farm because she and her husband had lived and farmed on the west end. She told about working there in a field in the woods.

> Down in the woods. I'd be so scareidy in the day. [Laugh]. . . . I'd be so scared something would catch me.

> [Laugh]. . . . It was good land. You get down there . . . I'd think something, a varmunk or something, going to come up out the swamp. . . . I thought a varmunk or something going to catch me before I come back out there. . . . But it was good land. . . .
>
> When we used to come and get hands from up here [in Florence] to go help and pick cotton, they'd tell about how that place was before I got there. But . . . all them gone. . . .
>
> They'd tell me about how they used to work back there and how low the land used to be in them days. . . . But they never did say about the rice. And they were surprised to see how we clean it up. And they say, "Who would have thought I'd be living to come help you pick cotton in this field when it used to be nothing but pawn. . . ." We called it pawn . . . a lot of lowland.[2]

Those were Mrs. Waiters' memories of fields in the woods. She had not solved the mystery of who cultivated the rice fields on the Broadwell place, so I did not include those fields in the tally of African American rice fields at Mars Bluff. Still, the location and the size of the two small fields indicate that they probably were African American.

Hester Waiters had a rice story of her own to tell. It came from her childhood days, when she had lived in the neck of land between the Great Pee Dee River and Jefferies Creek. She told how she had helped her stepfather, Horace Gregg, grow dryland rice. They had only grown it for one year, but she recalled that experience in great detail. She also recalled that on Saturdays she would carry a sack of rice to her uncle Benny's house to use his mortar and pestle. Uncle Benny was her father's brother, Benny Bailey, who lived nearby. Mrs. Waiters said that he grew rice regularly.[3]

To learn about Benny Bailey's rice field, I went to the neck—past Bailey's old farm and past three deer and one red fox. I found Bailey's son, Willie H. Bailey, a robust and industrious man, working in his tobacco field. He was reluctant to stop. His father had taught him to work in the morning and stop work in the heat of the day. It was too early to take a rest, so we stood in the tobacco field while he reached way back in his memory.

He said he was born about 1910, and he recalled that when he was a child his father grew a small spot of rice in a dryland field.

Fig. 35 No photograph of Hester Waiters exists, but here is the lane she walked as a child, carrying her sack of rice to her uncle Benny's house.

He would pick "a damp place because rice likes that better than a dry place," and he raised enough for his own family and "to give the neighbor a little." He had a mortar and pestle and did much of the husking himself.

Willie Bailey recalled that there were other rice growers in "that settlement"; one of them was named Roager Williams. Most of these

Fig. 36 Reluctant to stop work, Willie Bailey stands in his tobacco field for an interview. His father, Benny Bailey, grew rice in the 1920s.

rice growers stopped planting about the same time, sometime in the 1920s. They had not told him why they quit; he supposed that rice became more plentiful.[4]

Mrs. Willie Bailey recalled that as a child she had heard the adults say that her mother's father, Ellison Hunter, raised rice. Hunter had also lived in the neck section of Mars Bluff.[5]

African American Rice Growers at Mars Bluff

Grower	Cultivation Method	Field Size	Year Started	Year Stopped	Reason Stopped
Bailey, Benny	dryland	"a small spot"	in 1920s	in 1920s	?
Brown, Tom	wetland	"one acre" or less	?	after 1921	died
Charles, Lewis	?	?	?	before 1920s	?
Coker, Preston	dryland	"a couple of acres"	*ca.* 1933	*ca.* 1935	?
Daniels, Arthur	dryland	?	?	in 1920s	moved
Ellison, Fanny	dryland	"rows beside garden"	? (Mars Bluff) *ca.* 1925 (Florence)	*ca.* 1925 *ca.* 1930s	moved poor health
Fleming, Frank	wetland	115′ × 240′	?	1920	moved
Gregg, Alex	wetland	"two and a half acres" or less	?	*ca.* 1920	old age
Gregg, Horace	dryland	"a low place in field"	*ca.* 1916	*ca.* 1916	?
Harrell, Alex	wetland	?	?	*ca.* 1920	moved
Hayes, Jim	wetland	?	?	before 1920	died
Hayes, Willie	wetland	?	?	before 1920	died

Howard, Tony, and Robert Gregg	wetland	60′ × 300′	?	*ca*. 1920 or 1922	?
Hunter, Ellison	?	?	?	before 1930	?
[last name unknown] Jim	wetland	"one-half acre"	?	*ca*. 1920	died
Johnson, Vico	dryland	120′ × 150′	?	*ca*. 1939	?
Robinson, Alfred	wetland	?	?	after 1923 and before 1927	?
Saunders, Lawrence	wetland	"wasn't big"	by 1921	in 1920s	?
Scott, Willie	wetland	120′ × 120′	?	1928	died
Smalls, Lewis	dryland	"about a half acre"	*ca*. 1921	*ca*. 1923	?
Williams, Roager	?	?	?	in 1920s	?
Williamson, Sherman	wetland	60′ × 162′	*ca*. 1922	*ca*. 1923	?

Totals	
Wetland rice growers	12
Dryland rice growers	7
Method unknown	3
Total	22

There is solid evidence that 21 of the 22 grew rice. There is only circumstantial evidence that Lewis Charles grew rice.

In addition to the above figures, the 3 rice fields on the Broadwell farm were not included in this table because there was insufficient information about who had cultivated the fields.

Of the 12 wetland rice growers, 3 were known to have used water from a ditch for irrigation. The others were believed to have relied entirely on rain.

I had set out to learn from Mrs. Waiters who grew rice on the Broadwell farm. Instead, I had found four more rice growers in the southeast part of Mars Bluff.

Rice growers seemed to be everywhere. Already people had told me about twenty Mars Bluff African Americans who had raised rice, and later I would learn of two more. Those rice growers were found, not by any systematic survey, but largely through interviews with older African Americans about other topics. There certainly seemed to be a widespread knowledge of rice cultivation among African Americans at Mars Bluff.

9
Another Mystery

LACY RANKIN HARWELL, a European American who lived at the north edge of Mars Bluff, always took a lively interest in conversation, so visiting with her was a treat, and I frequently paid a call on her when I went to Mars Bluff.

On one visit we were talking about rice, and Mrs. Harwell said that Arthur Daniels, an African American who had lived on her farm, had grown nonirrigated rice until about 1930.[1] I was pleased to have found another rice grower, so I phoned Daniels' daughter to learn more about his field.

Isabell Daniels Smith described the field as lying at the base of the hill where the high land drops off into Back Swamp. She said that the field was not covered with water, but was a "dark, stiff, wet piece of land," and it produced just enough rice for their family.

Mrs. Smith recalled working with a grub hoe to cultivate the rice and bringing the sheaves of rice up the hill to the barn. Her father hung the sheaves on nails that projected from the ceiling joists. When they had dried, he would wring the heads off and then beat them on a sheet on the barn floor, using "something like a maul."[2] He stopped planting rice when he moved, in the 1920s.[3]

I appreciated Mrs. Harwell's telling me about Daniels, and I was even more grateful when she shared what she knew about the old irrigated rice fields on her farm. Although she had few facts, she knew more than anyone else.

Mrs. Harwell's home and dryland fields were on the same elevation as the rest of Mars Bluff; but immediately behind her house, a steep hill dropped down into Back Swamp. There at the base of the hill were the only remains of irrigated rice fields known to exist in Mars Bluff.

Mrs. Harwell recalled that when she was a child, she would go down the hill behind her house and go fishing off the causeway. The causeway was the road that crossed the first canal at the bottom of the hill. Beyond that lay the abandoned rice fields.

That was as far back as Mrs. Harwell's recollections went, to about 1910. She did not know anything about how the fields had been cultivated because irrigated rice cultivation there had ended so long before her time. No one knew exactly when it started or when it ended.

Mrs. Harwell did know a little about the earlier owners of the land. She told me about Abraham Satur, a Charleston merchant, who had a large grant of land in 1736, but his grant had lapsed and the land was sold in smaller parcels. She told me the names of some of those owners—and best of all, she gave me permission to go see the old fields.[4]

Of course I took advantage of that permission. I crossed the Harwells' dryland fields and went down the steep hill into Back Swamp. It was a paradise for a lover of black water. From an embankment east of the fields I looked down into a beautiful black-water canal and out across old fields now covered by a sparse growth of trees—and lots of black water.[5]

Following the embankment down the east side of the fields, I came to a place known to local hunters as "the old floodgate."[6] Nothing remained of the old gate except an opening in the embankment. Beyond the opening, a black-water canal led away from the rice fields. Those embankments, canals, and fields represented a tremendous amount of labor—they covered an area perhaps half a mile long and several hundred yards wide. I wanted to know when all that work had been done and when the fields were last used, so I began a search for information about the former owners of the Harwell land. Searching the records, I found the names of many of the early settlers; however, the deeds, plats, and probate records gave no clue about the rice fields.[7]

In 1844, the record became a little clearer. That year, Dr. William R. Johnson bought 1,631 acres of land, on which he lived until his death in 1893.[8] I thought that I would be able to establish that he was the last person to use the commercial rice fields.

The first clue I had was a little knowledge of Dr. Johnson's personality. He was an energetic and active man—he built a house that dwarfed

Fig. 37 This is what remains of one of the canals near the center of the old irrigated rice fields on the Harwell farm at Mars Bluff.

all other houses at Mars Bluff, and he rode horseback until his death as a very old man.[9] My hunch, based on Dr. Johnson's personality, was that whatever rice production was occurring when he acquired the land may well have been expanded during his tenure.

There was another clue. An article in the *Rice Journal* in 1920 was accompanied by a picture of the house that Dr. Johnson built in 1857.

Fig. 38 Johnson-Rankin-Harwell home (1857), located on the only farm at Mars Bluff known to have had irrigated rice fields.

Illustration taken from the Rice Journal, *XXIII (September, 1920), 27. Reproduced by permission of the* Rice Journal, *Raleigh, N.C.*

The picture was captioned "Residence of an ante-bellum Rice Planter."[10] So Dr. Johnson had been a rice planter. All I had to do, I thought, was find out what year he stopped planting rice.

But then Marion Chandler, an archivist at the South Carolina Department of Archives, showed me a piece of conflicting evidence. The agricultural census for 1850 itemized Dr. Johnson's farming operation for that year as follows:

Improved land, acres	600
Unimproved land, acres	1,700
Cash value of farm	20,000
Value of farming implements and machinery	500
Horses	8
Asses and mules	7
Milch cows	24
Working oxen	4
Other cattle	100
Sheep	0

continued

Swine	130
Value of livestock	2,800
Wheat, bushels	0
Rye, bushels	100
Indian corn, bushels	1,500
Oats, bushels	500
Rice, pounds	0
Tobacco, pounds	0
Ginned cotton, 400-lb. bales	43
Wool, pounds	0
Peas and beans, bushels	800
Irish potatoes, bushels	10
Sweet potatoes, bushels	500
Butter, pounds	150
Cheese	0
Value of home-made manufactures	0
Value of animals slaughtered	720

Dr. Johnson raised no rice in 1850. He did raise forty-three bales of cotton. He was a cotton farmer, not a rice farmer.[11]

Perhaps Dr. Johnson had cultivated rice when he first bought the land in 1844. Horace Rudisill produced a document that seemed to suggest so. It was an easement that a neighbor gave in 1846 that allowed Dr. Johnson to build and maintain an eight-foot-wide ditch across the neighbor's land in Back Swamp.[12] On first seeing the easement I assumed that Dr. Johnson wanted to dig that canal to get better control of the water level in his rice fields. But if he was not growing rice in 1850 it seems more likely that he dug the canal to drain the area so that he could use it for dryland crops.

Dr. Johnson might have produced rice after 1850, but that seems unlikely for two reasons: He was listed as producing no rice in the 1860 census, and Charlie Grant gave an account of life on the Johnson farm in the 1850s with no mention of rice. Grant, in a 1937 Federal Writers' Project interview, told about his life during the last years of slavery. Grant was quite young at that time, and his work was tending the sheep and cows. Still, it seems likely that he would have spoken of rice if rice was being cultivated there.[13] Altogether, the evidence seems to point to the conclusion that Dr. Johnson was never a rice planter, despite his description as such in the *Rice Journal* caption.

Does that mean that the use of the Harwell farm rice fields ended before 1850? Perhaps. But it is possible that in 1850 Dr. Johnson did not own the rice field land. His 1844 plat contains no landmarks except one small stretch of Black Creek, so it is impossible to tell exactly which land was his.[14] Maybe someone besides Dr. Johnson owned the rice fields.

The 1850 agricultural census showed that only seven people in Marion County had raised over ten thousand pounds of rice that year. One of those men—James C. Thompson, who raised twelve thousand pounds—may have been using part of the Harwell farm rice fields. One unsubstantial clue points in that direction. Mrs. Hannah Thompson owned land south of Dr. Johnson's land, and Mrs. Thompson's 1856 will speaks of "the estate of her deceased son, James C. Thompson." It is possible that Mrs. Thompson owned a portion of the rice fields and that her son was cultivating them.[15]

There is also weak evidence of a connection between the Harwell rice fields and Samuel Gibson, who raised forty thousand pounds of rice in 1850. It seems highly unlikely, however, that Gibson would have been using those fields.[16]

Perhaps no one had grown rice on the Harwell fields for decades before 1850. That would have been logical. As early as 1750, rice irrigated by the tidal-flow method had proved so profitable that inland rice farmers had gradually stopped trying to compete with growers along the coast.

Altogether, the search for facts about when the Harwell rice fields were first and last used had produced little compelling evidence. The mystery remained unsolved.

One mystery was solved—the source of the water that irrigated the fields. Had it come from Black Creek or from reserves? To answer that question, Timothy Dargan, a forester well acquainted with the swamp around the rice fields, drew a map of the fields based on a stereoscopic study of an aerial photograph. Included on the map were the two canals that led away from the rice fields toward Black Creek. The mapping of the canals demonstrated that they had not extended to the creek. That seemed fairly conclusive evidence that the fields did not get their water supply from the creek, leaving only one possibility: the irrigation system for the fields must have relied on reserves, reservoirs of water gathered from the surrounding area.[17]

Although the Harwell fields were the only remaining evidence of

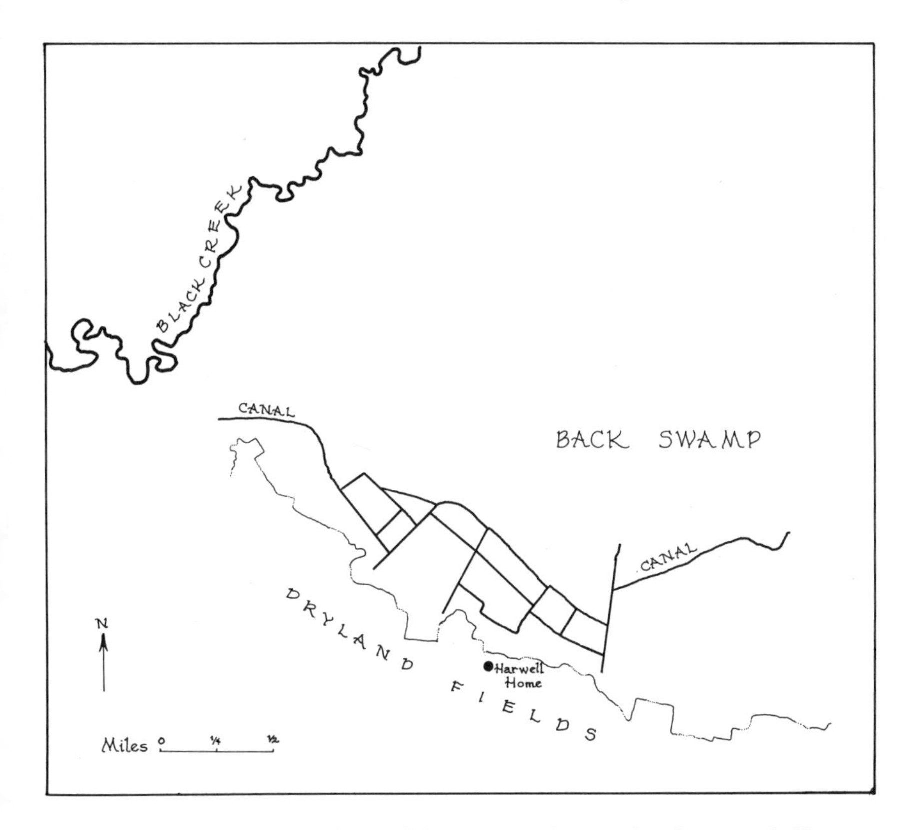

Map 7 Canals at irrigated rice fields, two miles north of Mars Bluff center. The Great Pee Dee River is several miles off to the east. The canals do not reach Black Creek or the Great Pee Dee. Timothy G. Dargan mapped the canals by stereoscopic study of government aerial photographs in 1988.

Map by Dinah Bervin Kerksieck.

irrigated rice fields at Mars Bluff, there was a clue that there may have been other irrigated fields in the area. Alex Gregg had said that three places needed help with the rice harvest. Here is how the conversation went when I asked Archie Waiters if his grandfather had told him anything about European Americans raising rice:

A. Yes, he say some white people raised rice. He say about—He know three raised rice.

Q. Who was that?

A. He didn't tell me. They didn't raise rice on the place they was living on.

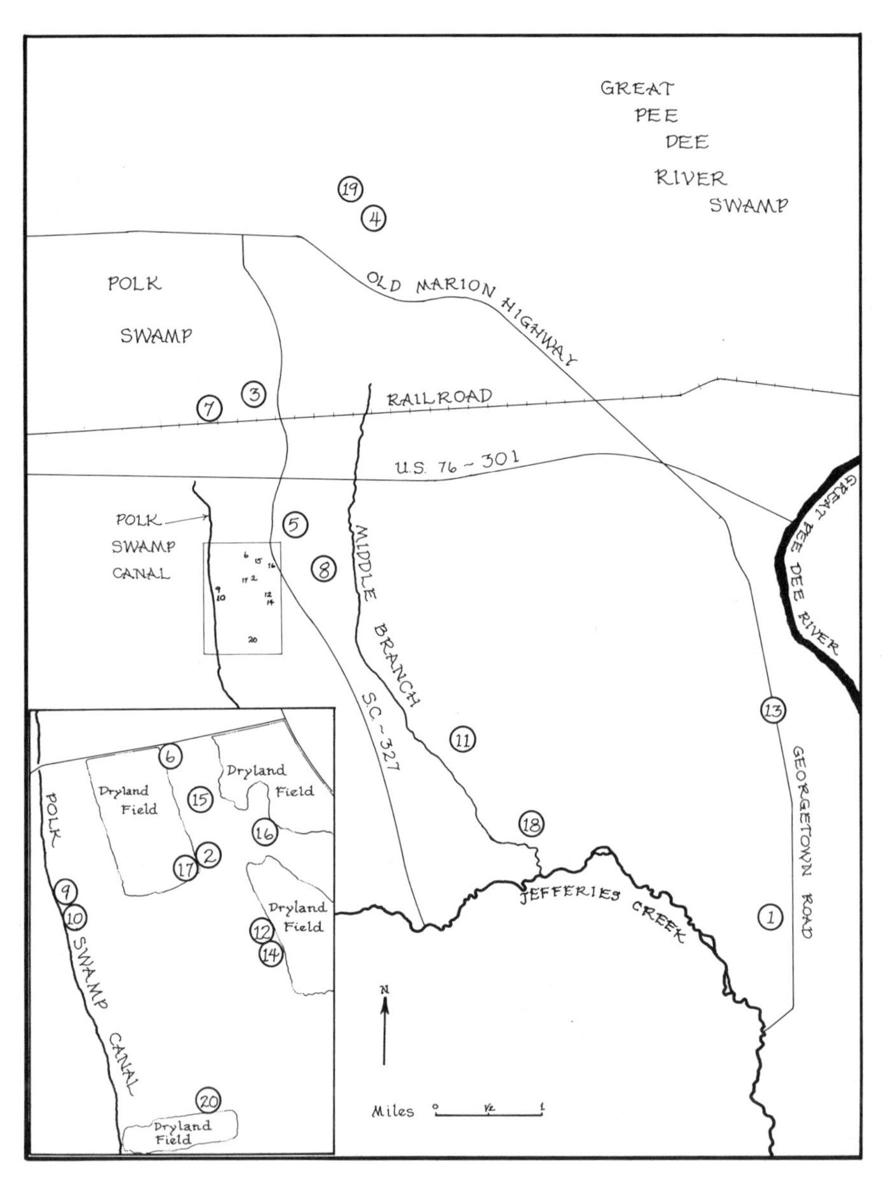

Map 8 Rice fields at Mars Bluff. The rice field of Steven Taylor, who grew rice in the 1930s and 1940s, lies about one mile west of the western limit of this map.

Map by Dinah Bervin Kerksieck.

African Americans' Small Rice Fields

(18 of 22 field locations known)

1.	Benny Bailey	10.	Willie (?) Hayes
2.	Tom Brown	11.	Tony Howard
3.	Preston Coker	12.	Jim (last name unknown)
4.	Arthur Daniels	13.	Vico Johnson
5.	Fanny Ellison	14.	Alfred Robinson
6.	Frank Fleming	15.	Lawrence Saunders
7.	Alex Gregg	16.	Willie Scott
8.	Alex Harrell	17.	Lewis Smalls
9.	Jim (?) Hayes	18.	Sherman Williamson

European American Large Irrigated Rice Fields

19. Harwell farm rice fields

European American Nonirrigated Rice Fields

20. J. Eli Gregg

Map 8 Key

> *Q.* But there were three people around here—
> *A.* Round in the neighborhood raised rice. . . . He tell me when they get ready to gather rice they come over there and get them to gather the rice for them.
> *Q.* How far would he have to go?
> *A.* I don't know how far he got to. Them people didn't mind walking about five or six miles. . . .
> *Q.* When he'd gather rice on this other man's place, he'd walk over there every morning?
> *A.* Walk over there every morning.
> *Q.* And walk back every evening?
> *A.* And walk back every evening.

Those large crops of rice that required extra help at harvest might have been made on unusually large nonirrigated fields, but more likely they would have been made on irrigated fields.[18]

While it will remain a mystery where Alex Gregg went to harvest rice, it was obvious from the evidence at the Harwell farm and from Alex Gregg's recollections that African Americans at Mars Bluff had contact with European Americans who were growing rice. So a question arose: Who learned what from whom about rice cultivation? That question deserved an answer. In fact, that question already had two answers.

10
The South Carolina Rice Story

WHO WAS IT that brought knowledge of rice cultivation methods to South Carolina? For years that question had been answered without mentioning African Americans. Here's how the story was told.

The first permanent settlers arrived at Charleston in 1670. Even before they arrived, the lords proprietors were looking for suitable crops for their new colony, especially for a cash crop that would lure settlers and make the proprietors rich. They thought of everything from wine and raisins to silk—and they had high hopes for rice.[1]

Just two years after the settlers arrived in Charleston, the proprietors sent them a barrel of rice, presumably for planting. No one knows for sure that any of it was planted, but it is possible that some people grew rice successfully at that time.[2]

One of the most often repeated stories about early rice cultivation in South Carolina involves a New England ship that was obliged to put into Charleston Harbor for repairs sometime before 1686. While there, the ship's captain, John Thurber, gave some Madagascar rice to Dr. Henry Woodward, a prominent citizen of Charleston. He "planted this seed, had a very good yield and distributed the seed among his friends for further planting." So the story is told that Woodward was the first person in South Carolina to grow Madagascar gold rice successfully.[3]

De Bow wrote of the earliest days of rice cultivation: "At first, rice was cultivated on the high land, and on little spots of low ground, as they were met with here and there. These low grounds being found to agree better with the plant, the inland swamps were cleared for the purpose of extending the culture."[4] By 1700, Carolina was exporting quantities of rice, and the quality of Carolina rice was highly esteemed in Europe.[5]

The early rice fields, whether dryland or wetland, depended entirely on rainfall for water; consequently, the success of the crop was highly variable from year to year. Gradually, South Carolinians evolved the "reserve method" of irrigation. Herbert Ravenel Sass wrote: "In the early days of rice planting . . . the culture of the grain was confined to the "inland swamps"—that is, the forested swamps not necessarily immediately adjacent to the rivers. An abundant supply of water being needed . . . was obtained by damming a swamp or lead situated at a slightly higher level than the fields, thus forming a lake known as a "reserve" or "backwater." From this reserve the water was drawn at need for use on the planted land."[6]

In times of drought the reserve would not supply enough water for the rice, so the more efficient tidal-flow method of irrigation was developed in the eighteenth century. It required that rice fields be laid out in swamps adjacent to rivers and close to the coast so that the level of fresh water in the rivers was affected by the tides. Massive embankments protected the fields. Floodgates in the banks could be opened to flood the fields with the fresh water that was backed up in the rivers by the high tide.[7]

The tidal-flow method of irrigation revolutionized commercial rice production, for it guaranteed a dependable water supply. But because the use of the tides for irrigation was effective only within about twenty-five or thirty miles of the coast, the area of commercial rice cultivation became largely limited to the narrow tidewater area.[8]

Those coastal plantations supplied rice to Europe and the rest of America until the twentieth century, when the rice market was captured by the mechanized rice farms of Louisiana. Although South Carolinians no longer produced much rice, they continued writing about the history of Carolina rice—telling and retelling the story. In all of those accounts, African Americans were never mentioned as having contributed anything but labor. No one suggested that Africans might have brought knowledge of rice cultivation from Africa to South Carolina.

The passage of time did not result in a more accurate account of the part played by African Americans. On the contrary, the early twentieth century saw an institutionalization of racism, and historians generally were silent about African Americans or portrayed them as savages with nothing to contribute.[9] Ulrich B. Phillips, who was esteemed as the foremost writer about slavery in the first half of the twentieth century, wrote that the plantations had worked with success "only when the management had . . . allowed for the crudity of the labor." He saw the plantation system as a step in African Americans' "essentially slow process of transition from barbarism to civilization."[10]

That sort of thinking was widespread. W. E. B. Du Bois wrote that Columbia University historians viewed African Americans with "ridicule, contempt, or silence." Early twentieth-century writing about rice cultivation in South Carolina followed the same pattern. There was silence about any African American contribution of knowledge, and there was contempt for African American laborers.[11]

As recently as 1972, when David Coon made a thorough study of the development of rice cultivation in South Carolina, he saw African Americans as having contributed nothing except labor; there was no suggestion that they might have brought or added to the knowledge on which rice-growing practices were based.[12]

That was the prevailing attitude in 1974 when a courageous young historian, Peter Wood, came upon the scene. Wood's book *Black Majority* challenged the way that the Carolina rice story was being told. He wrote: "Scholars have traditionally implied that African laborers were generally 'unskilled' and that this characteristic was particularly appropriate to the tedious work of rice cultivation. It may well be that something closer to the reverse was true early in South Carolina's development. . . . If highly specialized workers were not required, at the same time there was hardly a premium on being unskilled. . . . With respect to rice cultivation, particular know-how, rather than lack of it, was one factor which made black labor attractive to the English colonists."[13]

Wood found evidence that African Americans who came to the colony in the earliest days were appreciated for their versatility and competence in a number of areas. Their intelligence was acknowledged. "It is only from the more closed society of later times, which placed a high premium upon fostering ignorance and dependence within the servile labor force, that white Americans have derived the false notion that black slaves were initially accepted, and even sought, as being totally 'unskilled.'"[14]

Wood demonstrated that early English settlers in Carolina knew nothing about rice. In contrast, many Africans were experienced rice growers, and the European Americans in Carolina, according to Wood, made a special effort to secure Africans who were familiar with rice cultivation.[15]

Wood also showed that African Americans had been present in

Carolina from the very first year of settlement. Within two years they constituted a quarter of the population, and by 1708 they were a majority. "To a degree unique in American history they [African Americans] participated in—and in some ways dominated—the evolution" of the early Carolina rice country.[16]

Early documents suggest that it was those Africans who taught the English in Carolina how to grow rice. Wood wrote:

> Negro slaves, faced with limited food supplies before 1700 and encouraged to raise their own subsistence, could readily have succeeded in nurturing rice where their masters had failed. It would not have taken many such incidents to demonstrate to the anxious English that rice was a potential staple and that Africans were its most logical cultivators and processors. Some such chain of events appears entirely possible. If so, it could well have provided the background for Edward Randolph's comment of 1700, in his report to the Lords of Trade, that Englishmen in Carolina had "now found out the true way of raising and husking Rice."[17]

Other scholars accepted Wood's thesis. At last, African Americans were included as knowledgeable participants in the story of South Carolina rice cultivation.[18]

Then another young historian came upon the scene, Daniel C. Littlefield. He explored the question of which Africans knew about rice cultivation. Studying various peoples of Africa, he found that West Africans had rice-growing methods to suit every soil and water condition. He reported that there were "more varied systems and practices of rice cultivation in West Africa than in all Asia." He wrote:

> It is clear then that various African groups—Malinke, Soninke, Serer, Joola, Balante, Kisi, Papel, Baga, Mende, Temne, and others—were conversant with numerous varieties of rice, both African and Asian, and with various methods of rice cultivation, some of them quite advanced. It is also evident that knowledge of rice production was widespread in West Africa. . . . Even when West Africans made little attempt to alter the environment, their agricultural practices required a sophisticated knowledge of the terrain, soil types, and properties of

> different kinds of rice. In agricultural terms, therefore, the African was anything but an ignorant savage.[19]

These West Africans, with centuries of experience in raising rice, were quite capable of bringing knowledge of rice cultivation methods to Carolina. But were these the Africans who were brought to the New World?

By Littlefield's figures, 43 percent of the Africans brought into South Carolina during the eighteenth century came from the northern, rice-growing region. Forty percent came from the south, the Congo-Angola region. Only 17 percent came from the central part of Africa around the Niger delta. These figures stand in sharp contrast to those for other colonies like Virginia, to which the largest number of Africans came from the central part of the African coast. Fewer came from the south, and even fewer from the north.[20] Such disproportionate numbers showed clearly the South Carolina preference for Africans from the rice-growing region.

The work of Wood and Littlefield firmly established that Africans had made important contributions to Carolina rice cultivation—producing a culture that combined European and African elements.[21] Since then, many other scholars have added to an understanding of the African influence. Charles Joyner's study of African American culture in the tidewater rice region leaves no doubt that the Waccamaw rice plantations were indeed a merger of African and European cultures.[22]

Alex West's video "The Strength of These Arms: Black Labor—White Rice" includes 1940 footage of African Americans cultivating and processing rice by hand at White Hall Plantation on the Combahee River, in southern South Carolina. "Widespread commercial production of rice declined steadily after 1900," West said, "but the skills of processing rice by hand remained familiar throughout the low country as a means of subsistence."[23] It was that tradition of small-scale rice farming without irrigation that Mars Bluff rice growers had used. Their ancestors in Africa had raised rice for centuries, and they had carried knowledge of rice cultivation with them to Carolina. This was the knowledge that the Mars Bluff rice growers kept alive until well into the twentieth century.

11
The Mars Bluff Rice Story

HOW DID AFRICAN Americans in Mars Bluff maintain their African knowledge of rice cultivation as late as the 1920s? Slaves at Mars Bluff would have been obliged to work in the landowners' fields from sunrise to sunset, six days a week. They would have had little leisure time for working in gardens of their own—and less time for growing rice. As generations passed, it seems they would have forgotten their rice cultivation skills.

The situation was different in the tidewater region. There, slaves worked tasks. Each day, each person was assigned a task—a specific part of a field to plant or hoe. When the task was finished, the person was free for the rest of the day; so there was leisure time for fast workers.

R. Q. Mallard, writing about life on a tidewater rice plantation in Georgia, described the food ration given to each slave. He said, "To this, thrifty servants added rice, of which they were as fond as the Chinese." He explained that the industrious laborer in the irrigated rice fields could finish his task early in the day and be "ready to work his own allotted patch in the rice field."[1]

At Mars Bluff, the slaves worked in gangs, laboring together in the landowners' fields as long as there was daylight. There was no quitting early to go tend their own plots of rice, so very few slaves would have been able to grow rice for themselves. Under these circumstances, how could the slaves have kept alive their African knowledge of rice cultivation? The answer to that question may be found in the papers of European American landowners in the pine belt. Their journals and estate inventories, now yellow with age, show that small fields of rice were a part of their routine farming operations.[2]

The estate of John Fraser, an early settler at Mars Bluff, was inventoried in 1772. The record shows that indigo and corn were his major crops, but that rice was being produced in small quantities. At his death, Fraser owned the following:

300 bushels of corn
Indigo not packed up £350
10 bushels of indigo seed
Ruff rice[3] £10

The appraisement of the estate of another Mars Bluff man, John Gregg, in 1839 showed a change in agricultural practices from Fraser's time, almost seventy years earlier. Gregg's inventory listed 120 bales of cotton but no indigo. Rice was still present. Among the grains itemized were 2,500 bushels of corn, 160 bushels of oats, 45 bushels of rye, 25 bushels of rice, and 5 bushels of wheat.[4]

Decades later, during the Civil War, when agricultural products were taxed in kind, another Mars Bluff man, Dr. Robert Harllee, listed his crops. Cotton was still king; his food crops included 2,600 bushels of corn, 15 bushels of wheat, 100 bushels of rice, and 500 bushels of sweet potatoes.[5]

Further documentation of rice cultivation on land owned by European Americans at Mars Bluff is seen in the records of another pine belt planter, John Pinckney, of Sumter County. He drew up a plan for planting in 1865 that called for a small amount of rice:

PLAN OF CROP FOR THE YEAR 1865

	Acres
Corn	414
Cotton	40
Millet	10
Potatoes	25
Early peas	25
Rice and sundries	9

Pinckney wrote a revised plan for the same year, entitled "Plan of Crop, April 24, 1865—After Yankee Raid." He had reduced the amount of corn to be planted by about 100 acres and had eliminated cotton, but his little rice crop was unchanged: "Rice and Sundries, in small places—say 9 acres."[6]

All four of these pine belt European Americans claimed the rice crop as their own. It was in their estate inventories, their tax records, their plans for planting. But it was not the European Americans who were planting rice. Fraser owned 17 slaves; Gregg, 76; Harllee, 116; and Pinckney, 40.[7] It was the African Americans who were doing the actual growing of the rice—putting into practice their African knowledge of rice cultivation.[8]

Henry Davis, one of the rare people to write about small-scale rice cultivation in Florence County, maintained that "probably only a small quantity of the rice produced on Florence County lands passed into commerce, but undoubtedly a sufficient amount was raised to supply all local needs. It constituted a large part of the diet of the people of this section during the era of the Revolutionary War. No evidence of the existence of rice mills of that era in this section can now be found: but every plantation had its rice mortar and pestle."[9]

The widespread practice of farms raising their own rice had kept alive the knowledge of rice cultivation among African Americans at Mars Bluff. Through all their years in slavery, they were required to grow rice on land that belonged to the European American, on time that was claimed by the European American, and for the benefit of the European American. Of course the European American considered the rice crop his. He would write in his journal, "Planted rice today," as if *he* had planted rice that day.[10] All the while, the African Americans were the ones who were planting the rice, caring for it, harvesting it, threshing it, and husking it—using a body of knowledge about rice cultivation that had been brought from Africa. Although the prevailing ethos did not acknowledge that they possessed any heritage, they were keeping alive their African heritage of small-scale rice cultivation.[11]

In Africa, the people had grown small plots of subsistence rice, just enough to feed themselves. In the pine belt, it suited the landowners very well for the slaves to continue their practice of subsistence rice cultivation—but not for the slaves' subsistence. The owners wanted the rice for themselves. Slaves were fed cheaper food. Bushel for bushel, corn was several times less expensive to produce than rice; therefore, enslaved African Americans were generally fed corn.[12] Rice was reserved for European Americans to eat. Consequently, African Americans maintained their African knowledge of rice cultivation long after they had ceased to be rice eaters.

In fact, at Mars Bluff, there came to be a reversal in the eating habits of African Americans and European Americans. In Africa, many rice growers ate rice every day. Joseph Opala quoted people in Sierra Leone who boasted: "We're rice eaters. Until we've had rice, we don't feel like we've eaten."[13] At Mars Bluff, slaves were rarely fed rice. Three times a day they ate corn, in the form of mush, grits, or cornbread. On the other hand, European Americans adopted the practice of eating rice everyday for dinner. It became such a well-established habit that European Americans came to feel that they had not eaten dinner unless they had eaten rice.[14]

The landowners' desire for rice for themselves had sustained the African knowledge of rice cultivation.[15] But why, then, were there so many little rice fields concentrated at the center of Mars Bluff on J. Eli Gregg's land?

Archie Waiters answered that question when he recalled what his uncle Sidney and his uncle Bubba had told him. Their memories reached back to the 1880s. They recalled working in J. Eli Gregg's rice field in Polk Swamp, well south of Jim's rice field and set back hundreds of feet from the canal. Waiters said:

> They used to help him [J. Eli Gregg] work it back there in rice and stuff like that. Them is the onliest two tell me about working in rice. And used to help him gather the rice and used to haul rice to Darlington, Hartsville, and another little place off somewhere in Florence County. I can't call the name of it. Use to carry rice to them towns to sell in the wagon. . . .
>
> Carry the mule and wagon and get there and carry it around to each store and deliver it out. See, they send two or three loads of rice over there in sacks and deliver it out to them stores. Like you go to Darlington in your wagon, I go to Hartsville in my wagon. The other wagon go to the other town.[16]

On another occasion Waiters was telling about African Americans who raised rice in the twentieth century. When I asked if they were paid for it, he said: "Yes . . . sure they get money for that rice. . . . I know he [the manager of J. Eli Gregg and Son] had to pay them, because I know he sell it, because he send the wagon and send two men with every wagon. Because sometimes he had six wagons, five wagons, seven wagons going. . . . three or four turns on the wagons."[17]

There were many small rice fields at the center of Mars Bluff because the landowner was a merchant. He not only wanted rice for his own family to eat; he also wanted quantities of rice to sell in nearby towns. African Americans were growing rice to satisfy the merchant's need.

Before emancipation, the landowner-merchant had required the slaves to grow rice for his family and his store. After emancipation, he wanted the freedmen to continue working his fields just as they had in slavery except that they would be paid wages. Cotton and corn were well suited to the old system of gang labor, but rice plots were small—hardly places for gangs to work. Landowners may have seen the advantage of letting freedmen cultivate rice independently; consequently, African Americans may have found in rice one of their few opportunities to be independent farmers and to receive cash for their crop.[18]

Joe Opala saw the Mars Bluff rice growers as people carrying on a proud tradition. Because growing rice requires more knowledge and skill than the cultivation of most other crops, Opala said that rice growers everywhere felt a sense of pride in their skill. It gave them status.

Opala thought that sense of status might explain why some Mars Bluff African Americans continued planting their small fields of rice into the 1920s, when there was apparently no merchant to buy their crop. They may have been reluctant to give up growing rice because it was a part of their way of life and they felt a deep sense of pride in their skill. That skill had set them apart as special, and it had given them a degree of autonomy when autonomy was a hard thing for an African American to find.[19]

Although the economic system at Mars Bluff changed over the years, one thing remained constant. From the time of their arrival at Mars Bluff to the first part of the twentieth century, African Americans had used their African knowledge of rice cultivation continually. During those years, the tidewater rice plantations had become immensely wealthy, then fallen into decline, and finally ceased to exist. Still, the Mars Bluff African Americans went on planting their small fields of rice.

Such tenacity in the maintenance of this custom reflected the resilience of the people's culture. Although much had been lost, the African heritage could still be clearly seen in the Mars Bluff rice growers.

Part Three: *Tom Brown*

12
Tom Brown's Riddle

I NEVER KNEW Tom Brown; he died in the 1920s. Still, I felt that I knew him from Archie Waiters' stories, and I always had a kindly feeling for the old man. But now I was cross with him, for just when I thought that I understood the Mars Bluff rice growers, Waiters told me two things about Tom Brown that did not fit into the picture.

The two unusual things about Brown were his diet and his interest in clearing land. Until I could make sense of those, I knew that I had not fully understood the rice story.

Waiters was talking about Brown, and I asked:

Q. Was Tom Brown a big man?

A. Great Da! He weighed around two hundred pounds.

Q. Was he tall?

A. Tall, just about as tall as I is.

Q. And he was in good health right up to the end?

A. All of them. All those old people in good health. Was in good health until he died.

Q. What did they eat that kept them so healthy?

A. Just what I tell you. Just the grits and didn't worry all that more meat. . . . Didn't eat no meat. Mush. All of them had cows. Grits and stuff like that. Didn't eat much rice.

Q. They didn't eat much rice?

A. No.

Q. How come?

A. I don't know why they didn't eat much rice, but they didn't eat much rice.

Q. They grew it?
A. They grew it, but they didn't eat it.
Q. When would they eat it?
A. Just like you come up here. Like you want to visit, to visit here. We come eat. You don't eat [the rice], leave a lot, will be throw it to the chicken or the hogs.
Q. Say that again?
A. What you don't eat will be throw out to the chicken—[out] the door, with the hogs.
Q. After you cooked it?
A. After you cook it—that what you don't eat be—what you don't eat be gone down the drain.
Q. So they cook it regular?
A. No, they don't cook it regular. They cook—just like you come here, think you love rice, they cook rice. Then what leave, the chicken or the dog or the hog get that.
Q. Yeah?
A. That's right. That what they do. They eat that big hominy. Call it big hominy. Take that and fry it, fry piece of fatback meat. And put it in that grease and stir it and fry it, and let it kind of burn. Put a little bit of salt and black pepper. Man. . . . That thing make you swallow your tongue. . . .
Q. Browns didn't eat a lot of rice?
A. No, didn't eat a lot of rice.
Q. But they grew a lot?
A. Grew a lot. But they didn't eat a lot of rice.
Q. Would they eat it on Sunday?
A. No . . . they didn't hardly eat rice. Just like you go visiting; they would cook rice like you want it, and that what you don't eat, throw that thing to the dogs or to the hogs. . . . They eat that big hominy. That why they stay so healthy. They didn't eat all that kind of greasy and heavy rations, too rich and stuff like that. Milk. That buttermilk. They cook mush. And piece of fatback.[1]

Here Waiters was saying that old people like the Browns always served rice to a guest but did not eat it themselves. Waiters explained that people the age of his grandfather and Tom Brown preferred to eat mush, grits, and big hominy, although they grew rice.[2] Many of their ancestors had probably come from rice-growing regions in Africa and had probably considered rice their primary food. Why didn't Tom Brown and Alex Gregg eat rice?

We have already noted that the slaves of Mars Bluff usually ate corn products three times a day. That was a fairly universal custom throughout the South. Leland Ferguson explains, "In the southern colonies native Indian corn or maize, well adapted to the hot, humid climate, became the primary foodstuff of American slaves, even in the 'Rice Kingdom' of Carolina and Georgia. To planters, rice was a valuable cash crop, whereas corn was relatively cheap, suitable food for slaves."[3] Even after emancipation, corn was still the staple in the African American's diet. Archie Waiters told of the monthly ration for farm workers at Mars Bluff in the early twentieth century: "a piece of meat, a bushel of [corn]meal, and a gallon of black moriah molasses."[4]

The habitual eating of corn over many generations may help explain why Brown and his wife preferred corn to rice. Anyone who has ever cooked a pot of one kind of cereal for a family that is in the habit of eating another kind of cereal can testify to the fact that, at the conclusion of the meal, there remains a whole pot of uneaten cereal. People prefer the food that they are accustomed to eating.[5] I would suppose that was one reason that the Browns preferred mush, big hominy, and cornbread to rice, even though their ancestors may have eaten rice in Africa.

The sum of the matter was, Tom Brown raised rice but did not like to eat it. That was curious, but Tom Brown's interest in clearing land was even more difficult to understand.

Waiters and I were talking, and I asked:

Q. How old was Tom Brown?
A. He was about my granddaddy age [born 1844]. . . .
Q. Did Tom Brown grow rice right up until he died?
A. Right up until he died [mid-1920s].
Q. Was he sick long before he died?
A. He take sick like today, had to knock off . . . and went home and he died before day that next morning.
Q. He was cleaning up land?
A. Cleaning up land.
Q. Where was it?
A. Back on his rice patch.
Q. Really? And he was cleaning up more land to plant more rice?
A. Plant more rice the next year. He was cleaning up. He take

sick. He knock off; he come home and he die before day that morning.

Q. And he was old as your granddaddy?

A. . . . he was old as my granddaddy. . . .

Q. Did he have any children?

A. No, he didn't have no children.[6]

It did not make any sense. Why would such an old man be interested in such a strenuous task in order to produce a little more rice when he didn't even like to eat rice?

Brown would not have been planting rice so that he would have more to sell. There probably was no market for his rice by 1920, for by that time Gregg and Son would no longer have sent wagonloads of rice to nearby towns. If Brown had wanted more cash-crop land, there was a more direct way to get it than by clearing land. As a sharecropper, he could have gotten more cleared land for cotton simply by telling the landowner that he could handle more acres.

Maybe Brown liked to grow rice because the landowner took no share of the rice crop. If Brown grew cotton, the landowner took two-thirds of the crop. But as Archie Waiters said, "Rice don't count because they give that to you."[7] Waiters was saying that African Americans who cleared worthless swampland for rice were permitted to use that land as if it were their own. They paid no share of the rice crop to the landowner.

But even that motivation seemed inadequate to explain Brown's strenuous land-clearing work. The old man was clearing land to plant more rice on the day before he died when he did not eat rice and could not sell it. I could make no sense out of Tom Brown's land clearing—until I talked to Rev. Frank Saunders.

13

The Reverend Frank Saunders

REV. FRANK SAUNDERS answered the telephone at his home in Florence. I was calling him to ask if he had a picture of his grandfather Frank Fleming. I never dreamed that Rev. Saunders knew anything about rice—and I certainly would never have guessed that he might help solve the mystery of why Tom Brown was clearing land for rice on the day before he died.

The first reply that Rev. Saunders made to my query was surprise enough. He said, "I don't have a picture of my grandfather; but my father, Lawrence Saunders, raised rice at Mars Bluff." Of course I wanted to see the field, so Rev. Saunders, a distinguished-looking man in his seventies, came to Mars Bluff to show me where it had been. He described his father as "kind of tall, weighed about 180, light brown, very handsome man. . . . His hair was almost like his mother's—black, long; it didn't take long for anyone to tell he was mixed with Indian."[1]

We set out to walk through the steamy woods to his father's old rice field, but it was too hot. We were about halfway there when we decided on a compromise: he would tell me how to find the field.

> *A.* Right down there. When you come out the end of this graveyard, if you get out the trees . . . maybe you can imagine a house [where he had lived as child]. . . . Behind that house, there was a little ditch. And between that ditch and this one here, that's where they had the farm, the little rice patch, because it was like the fellow say, it was wet down there. . . .
>
> When you cross the ditch you see that low place there. That's got to be it. . . .

The rice patch was about an acre long. Wasn't no big field. . . .

Q. How wide was it?

A. About, I'd say, about a half an acre. There wasn't a whole lot of it. . . .

Q. What years would your father have been planting that field? . . .

A. I was quite a boy when he start planting. [This would have been in the 1920s.] Looks like . . . it just kind of new happened, see. I don't know where the idea came from, but they say they going to plant some rice. And they planted and made it.

Q. And you say your father helped your grandfather clean up the land?

A. They mostly work together because there was pretty-good-size families in most of my granddaddy's sons. . . . My daddy sharecropped, as well as my granddaddy. . . .

Q. Was the field wet?

A. Yes. It was wet. It was wet.

At first . . . they dug up a place. They work hard for that thing. Dug it up, the roots and everything. Drag it out just like you do a tobacco bed. And got it fix off just as soft as anything. What happened, they made some land. Wasn't nothing but woods and they made land out of it. And I remember that, because they used to have that thing with knots burning down there in the night and everything. Digging it up and plant that rice. I don't know what I would have done—if they was giving me the land, I don't think I would do it. But I think they enjoyed [it]. . . .

. . . Those folks dig, cut, drag down some of everything and when they got through with this thing, it was a narrow spot of land that they had made.[2]

Rev. Saunders spoke with such enthusiasm that I could almost see the people working to clear the forest. I could smell the pine knots burning in the night—and I could feel the joy of "making land." In fact, Saunders spoke with such fervor about making land that I saw the Mars Bluff rice growers in an entirely new light.

Now I saw, for the first time, African Americans on their own initiative, joyfully undertaking this backbreaking and time-consuming work—clearing a forest on someone else's land without any pay for their labor. And what is more, they felt tremendous delight in the work and pride in their accomplishment. Why would they have undertaken such an arduous task so cheerfully?

I could not understand it; I only understood that Rev. Saunders' story had made rice production seem to be secondary. What was more important was the fact that they were making land. If that was true, then the old man, Tom Brown, might have been clearing land not so that he could produce more rice the next year but just for the sake of clearing land. But why would people find such a sense of joy and accomplishment in clearing land?

They may have enjoyed land clearing simply because it was a longstanding African tradition. For centuries Africans had been forced to clear land regularly because African soil had minimal fertility. In the tropical climate there was no buildup of soil fertility. High temperatures, heavy seasonal rains, and large amounts of sunlight all worked together to deplete the soil of nutrients; consequently, there was a regular need to clear new ground.[3] Was it this enduring African tradition that African Americans were recalling in their land clearing? That hardly seemed an adequate reason to undertake such exhausting work.

Maybe a better explanation can be found in Herskovits. He wrote that the most tenaciously retained part of a culture is its central focus, and the central focus of the African culture is religious and spiritual.[4] "The materials everywhere underscore the fact that religion gives color and coherence to these cultures of Africa. . . . To study agriculture without taking into account its supernatural sanctions as these bear on beliefs concerning ancestors and gods, or to investigate African social organization, or property rights, without probing the role of the ancestral cult is to obtain and present a truncated version of the phenomena being studied."[5]

I had tried to understand Tom Brown's land-clearing practices without taking into account the spirituality permeating African American life, and I had not understood how the West African felt about the land. G. Howard Jones wrote: "The West African has a peculiarly wonderful conception of the land. . . . He worships it not only religiously, but also patriotically in the truest sense as the home of his ancestors and of his heirs, . . . and finally he regards it, not selfishly for his own use only in his lifetime, but as something entrusted to him by his ancestors and as something to be passed on unimpaired or improved to his children's children for ever." Both Herskovits and Jones saw clearly the spiritual element in all African farming activities. The African people felt a mystical connectedness between themselves, the earth, and their ancestors.[6]

The strong spiritual ties between the people of Africa and their ancestors extended even across the Atlantic to Africans who had been taken to the New World. Herskovits wrote of Felix, of Togoland, whose family had lost many members to slavers.

> The people who were taken away, Felix said, had never been heard from. His family did not know whether any of them had lived long enough to leave descendants. But these relatives, wherever they had died, were still members of the ancestral generations, and that is why today, when Felix's family in the city of Anecho in Togoland give food for the dead, they call also upon those who had died far away to come and partake of the offerings. . . .
>
> Our conversations with Felix and the Dahomeans led us to the knowledge of how much in Dahomey, at least, the slaving operations were remembered, and led us finally to the cult of the ancestors. Ancestor worship, one of the most important cults not alone in Dahomey, but in all of West Africa, takes as its typical ceremonial form the annual "customs" for the family dead. At these customs food is given to the spirits of the recent dead, and to those of all the other ancestral generations; to those who died at home, and to those who met their end in distant lands; to those whose names are known and remembered, to those who have been forgotten, and to those whose names the family in Dahomey had never had the opportunity to hear.

That sense of mystical union with the ancestors, so strong that it could reach across the Atlantic from Dahomey to unknown lands, surely must have reached back from people like Tom Brown to the ancestors, both in the New World and in Africa.[7]

Paul Richards, a longtime student of West Africa, felt a strong sense of African spiritual values in Tom Brown's land-clearing activities on the day before he died. Richards believed that the old man was acting in a way consistent with laying claim to a place where future generations might be rooted, and where, sensing perhaps that he was close to death, he might in due course be recognized as an ancestor.[8]

I thought I was beginning to understand Tom Brown. He was clearing land just before he died because he saw the world in African terms. He was participating in a mystical communion with the earth and the ancestors. But that did not explain why Rev. Saunders had spoken with such enthusiasm about "making land." I knew that I still had not fully understood.

14
Tom Brown's Land

TOM BROWN WAS a young man at the time of the Civil War and Reconstruction. That might have been one reason why, many years later, he was clearing land on the day before he died. He had African roots—but New World experiences had shaped his life.

During the tumultuous years of war and Reconstruction, Brown knew that his fate was being decided on the battlefields and in Washington, D.C. He would have been listening carefully for any news, especially for news about freedom and land.[1] A Union officer wrote about the hope of South Carolina African Americans at that time. This no doubt was Tom Brown's hope. "The great mass of Negroes in South Carolina at the end of the Civil War hoped and expected that freedom meant that each would soon be settled upon his own plot of earth. Indeed, to the negro agrarian freedom without land was incomprehensible. 'Gib us our own land and we take care ourselves, but widout land, de ole massas can hire us or starve us, as dey please.'"[2]

Wendell Phillips said the same thing in different words: "Before we leave him [the African American] we ought to leave him on his own soil, in his own house, with the right to the ballot and the school-house within reach."[3]

"On his own soil, in his own house." That was the key to real freedom. African Americans had come from an African tradition of subsistence farming. All they needed was to own a small plot of land where they could build a house and raise a garden, providing shelter and food for themselves. Then they would be free. They would be free to offer themselves in a free labor market. Until they had that minimum economic freedom, they were not really emancipated.

With a little more land, each family could raise a cash crop and be completely independent, but that was not essential. The one thing that was necessary was ownership of a tiny plot for a house and garden. Then African Americans would be freedmen in fact as well as in name. That was their dream and their hope.[4]

One might view the African American dream of land ownership as a naïve expectation with no basis in fact. Actually, it had a solid foundation in the events of the 1860s. Both during and after the war, the United States government gave African Americans reason to believe that their need for land was understood and that land would be made available to them.[5]

Early in the war, Sea Island plantations in South Carolina had been seized by the federal government on the grounds of abandonment and failure to pay the federal tax. In 1863, President Lincoln gave the first order indicating that it was government policy to help African Americans acquire land. He directed the tax commissioners to sell some of the seized land to African Americans in twenty-acre tracts at a preferred rate of $1.25 an acre. Of course African Americans were overjoyed. They staked out their plots of land. Their dearest hope, land ownership, was becoming a reality.[6]

Lincoln's directive set aside only twenty thousand acres for African Americans to purchase. General Rufus Saxton, the military governor responsible for the South Carolina Sea Islands, believed that the amount was inadequate. He thought that all African Americans who had toiled on the land for generations should be given an opportunity to purchase a small acreage. He persuaded the president to set aside more land for the freedmen, and hundreds of African Americans knew the joy of staking out the little farms that would be their homes.[7]

That sort of news would have traveled fast. No doubt Tom Brown heard the news and dreamed of the day when he would own a little piece of land. Unfortunately, the Sea Island tax commissioners had a different idea about the disposition of the land; they wanted to sell it in open bidding to raise revenue. The controversy raged for months. By the time the sale occurred early in 1864, the government had reversed its position; the tax commission had its way, and most of the land was put up for public sale. Many of the freedmen who had staked out small plots were driven off the land. A nightmare of disappointments and frustrated expectations had begun.[8]

In the closing months of the war, General William Sherman arrived at the coast with thousands of African Americans following his army. He consulted with the secretary of war and issued Special Field Order No. 15, which provided for the distribution of forty-acre tracts of coastal land to African Americans. The land involved was a strip thirty miles wide from Charleston to the St. John's River in Florida. Shortly thereafter, Sherman offered the loan of mules to the new landowners. This could be where the phrase "forty acres and a mule" originated.[9]

General Saxton offered his resignation rather than participate in another land-distribution scheme that would raise the hopes of the freedmen and then disappoint them. He was assured by the secretary of war that this time the arrangements would be permanent. The freedmen received similar assurances, and thousands of them settled on their own acreages.[10]

Surely Tom Brown heard of this. He must have felt that the time was near when he would be a free man—and would own a piece of land.

The government made a third commitment to help freedmen acquire land. In March, 1865, the United States Congress created the Bureau of Refugees, Freedmen, and Abandoned Land. One of the purposes of the bureau was to make land available for freedmen to rent in tracts of not more than forty acres; and those freedmen who were able could buy the tracts.[11]

When the war ended in April, 1865, thousands of freedmen were raising crops on their own small farms.[12] They had their own land, and they were free indeed. But their joy was short-lived. Within two months, President Andrew Johnson granted amnesty to many southerners. He ordered that those who had lost their land owing to abandonment and failure to pay taxes should have the land restored to them. In the fall of that year, the president ordered that the lands designated by General Sherman for distribution to African Americans be restored to their former owners.[13]

The Freedmen's Bureau tried to defend the rights of the freedmen. General Saxton wrote to the head of the bureau, General Oliver Howard:

> The lands which have been taken possession of by this bureau have been solemnly pledged to the freedmen. The law of Congress has been published to them, and all agents of this bureau acting under your order have provided lands to these freedmen. Thousands of them are already located on tracts of forty

> acres each. Their love of the soil and desire to own farms amounts to a passion—it appears to be the dearest hope of their lives. I sincerely trust the government will never break its faith with a single one of these colonists by driving him from the home which he has been promised.[14]

General Howard tried repeatedly to persuade President Johnson that the African Americans should be allowed to keep the acreages they were cultivating; but the president was adamant. In the end, Howard was forced to go to South Carolina to tell the freedmen that their farms must be restored to the former owners.[15]

Du Bois reported on one such meeting with freedmen in which General Howard, "to cover his own confusion and sympathy, asked them to sing. Immediately an old woman on the outskirts of the meeting began, 'Nobody Knows the Trouble I've Seen.' Howard wept."[16]

Edisto Island freedmen wrote to General Howard: "General, we want [h]omesteads. We were promised [h]omesteads by the government. If it does not carry out the promises [i]ts agents made to us . . . we are left in a more unpleasant condition than our former. We are at the mercy of those who are combined to prevent us from getting land enough to lay our [f]athers['] bones upon. . . . We are landless and [h]omeless. . . . We cannot resist . . . [w]ithout being driven out [h]omeless upon the road."[17]

Abandoned by the government in their claim to the land, the freedmen tried to buy land from the restored owners. They wrote President Johnson to tell him of the futility of their efforts: "Their late owners . . . had refused . . . to sell even so much as an acre and a half to each family, declaring that he would not take a hundred dollars an acre." It was clear that without government help the freedmen would not be able to acquire land. But the president did nothing to assist them.[18]

Congress made some ineffectual efforts to help the freedmen secure land. Although the Freedmen's Bureau had been established and one of its functions was to oversee the rental and sale of abandoned lands to freedmen, Congress would pass no law to make permanent the titles to the lands, nor would Congress make appropriations for the bureau's work.[19] A few legislators saw that government purchase of land was the only way to solve the problem. They introduced legislation, but all attempts to secure appropriations for the purchase of land failed because the plight of freedmen was not a primary concern of most congressmen.[20] They were more worried about how four million new citizens in the South would affect the balance of power in Congress.[21]

That was how the situation looked in Washington, but how did it look to Tom Brown in South Carolina? Of course Brown left no written record of his thoughts; however, John Dennett, a reporter who toured the South just after the war, was about twenty miles from Brown's home in October, 1865, when he interviewed some men whose thoughts were probably similar to Brown's at that time. Dennett wrote:

> Yesterday morning, just after leaving Marion village, I turned from the main road. . . . Very soon I lost my way . . . and rode aside to the Negro quarters to ask for information. . . I said to two of the Negroes: "Well, boys, have you raised much cotton and corn this year?"
>
> "Raise corn, sir. . . ."
>
> "You'll make cotton this next year, I suppose?"
>
> "Dunno, boss, dunno. We's waiting till Jenewerry come. Den we kin know."
>
> "Are you going to stay here another year?"
>
> "'Spect so, sir. Major G . . . says we kin stay ef we work, and we can have half the cotton we make."
>
> "But you won't make a contract till January, you say?"
>
> "No, sir," said the old man, "we heares dis an' dat, dis an' dat, an' we told him we'd hol' on tell Jenewerry. . . ."
>
> I waited . . . to see whether or not they would broach the subject of a division of lands among the Negroes, for I am told that such an opinion is universally prevalent in the lower districts of this State. But they also waited, and seemed disinclined to speak plainly, so I asked if that was what they were expecting to take place in January. "Yes," they said with some hesitation, "they'd been told so. . . ."
>
> When I told them that it was unlikely that any land would be given away by the Government, they listened to what I said, but appeared to receive it with dissatisfaction and incredulity.[22]

Freedmen in the area where Brown lived could not believe that the government was going to renege on its promise to help them get land. But the fact was that land distribution had already been ended by President Johnson's order to restore the land to the former owners. Late in

1865, the Freedmen's Bureau had to undertake the doleful task of persuading the freedmen that there would be no land distribution and that they must sign labor contracts for 1866.[23]

Six months later, however, Congress passed the Southern Homestead Act of 1866. It gave African Americans and European Americans who had not supported the Confederacy a six-month period in which they had first rights to settle on public land in five southern states. Here was a fresh hope for land.

But never was a hope so beset with impossible conditions. The freedmen were bound in labor contracts for that six-month period. With no food, tools, or capital, they had to move hundreds of miles away to poor, uncleared land. Frequently the homestead offices were not ready for business or had no officer available to sign the necessary papers. Few African Americans were able to overcome all of these obstacles. It was amazing that any succeeded in homesteading under these conditions in a region where African Americans were not wanted as landowners.[24]

The Southern Homestead Act of 1866 was the last attempt of the federal government to help freedmen acquire land, and most freedmen received no land—only dashed hopes. But now there was a new cause for hope in South Carolina. Congress had mandated that the state hold a constitutional convention; surely when the convention met early in 1868, it would make provisions for freedmen to acquire land.[25]

That was what South Carolina freedmen were hoping. But hope was one thing—what was actually happening to Tom Brown was something else altogether.

Walter Gregg, the owner of the land on which Brown lived, wrote to his wife on December 26, 1867: "I have not as yet gotten all the labor I wish for another year as I am trying to hire for wages; and a portion of the crop is the popular idea of the freedmen; but as labor will be abundant I have no fears that I will be able to accomplish what I wish."[26] What a revealing sentence. Gregg knew that he would get all the labor he needed—on his own terms. He did not even have to consent to the freedmen's request for sharecropping. Brown, who could neither buy nor rent land, would not even be allowed to sharecrop. He would have to work as a wage laborer doing gang labor in the landowner's fields, on the landowner's crops, just as he had before emancipation.

That was what was happening to Brown. But what was he thinking? Gregg wrote this letter less than a week before the beginning of the work year, and the freedmen still had not signed a labor contract. Gregg implied they were holding out because they wanted to sharecrop. True, but no doubt they were also hoping that the government would make good on its promise to provide them with land of their own.

Amazingly, the state constitutional convention of 1868 did act on the question of land for freedmen. Of all the states, South Carolina was the only one that attempted to make land available for freedmen to purchase. The convention directed the legislature to establish a land commission to buy land and resell it in parcels of twenty-five to one hundred acres. The freedmen had cause for rejoicing: their hopes had been well founded; the state government was going to offer them land for sale.[27]

The legislature established the commission and appropriated money to buy land. But then things went awry. Over the next three years, a corrupt advisory board embezzled land commission funds on a massive scale; the members of the board received large kickbacks from purchases; they bought worthless land to make political allies. Everyone was being paid off. Conditions became so bad that the advisory board paid the land commissioner $45,000 of land commission funds to get him to resign and keep silent about the corruption.[28]

Amid that scene of thievery, there were occasional men known for their honesty and their dedication to the welfare of the freedmen. One such man was Henry Hayne, an African American who was deputy land commissioner in Marion County, the county in which Tom Brown lived.[29] Through Hayne's efforts, Marion ranked fifth among the counties of South Carolina in acres of land acquired by the commission.[30]

Brown must have felt that his hope for land had not been in vain. The government was making good on its promise, helping freedmen to buy land—right there in Marion County. Twenty freedmen had bought one tract just across the river from Mars Bluff. Surely Mars Bluff freedmen rejoiced when they heard this, believing that they too would soon have land.[31]

The land commission continued in existence until 1889. For twenty years, knowledge of its existence fed the hopes of men like Brown. In its last years, however, the commission was used to make money for the state, selling land in very large tracts in violation of the law.[32]

About fourteen thousand families attempted purchase of land commission tracts; of these, probably about six thousand families acquired permanent title to land. It was a small number given the massive need that existed. This was the last effort by the government to provide land for freedmen, and it left most of them landless and discouraged.[33]

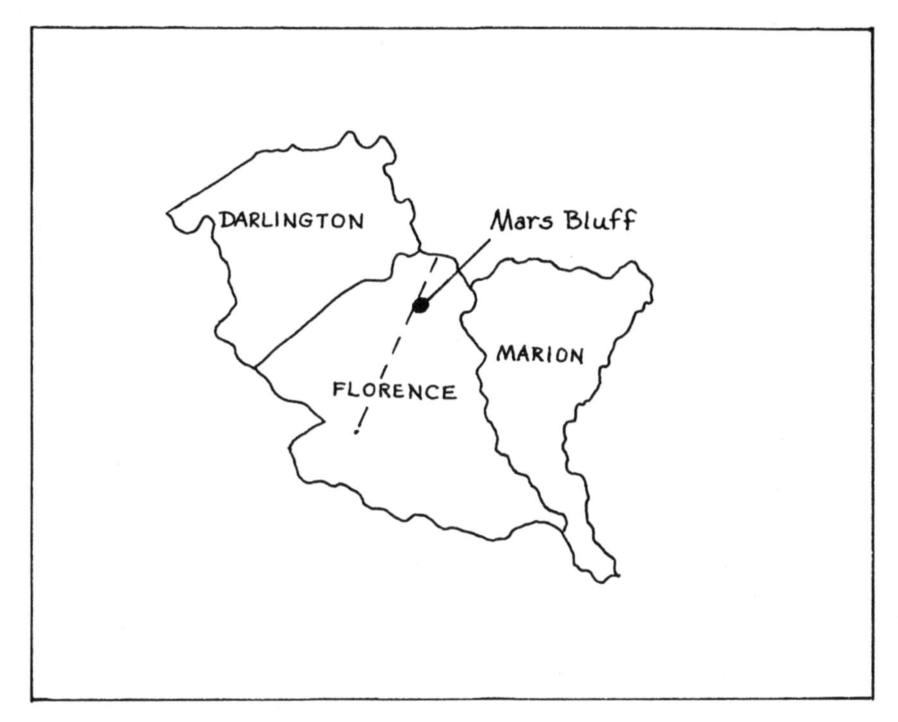

Map 9 County location of Mars Bluff. The dotted line is the old dividing line between Darlington and Marion counties. Mars Bluff was largely in Marion County, with the western tip in Darlington County. In 1888, Florence County was formed from parts of Darlington, Marion, Williamsburg, and Clarendon counties; and Mars Bluff became a part of the new county.

Map by Dinah Bervin Kerksieck.

Still, African Americans kept hoping. Phrases from Du Bois best describe the situation. "This land hunger—this absolutely fundamental and essential thing to any real emancipation of the slaves . . ." "For a long time there persisted the idea that the government was going to make a distribution of land." But the government never made any lasting provision for the freedman to acquire land. "In the end this spelled for him the continuation of slavery."[34]

Buying land directly from European American landowners was not a possibility either, for the owners adamantly refused to sell land to African Americans. They understood that if African Americans could be kept landless and dependent on landowners for their housing and their gardens, the landowners still had slaves. They had workers who

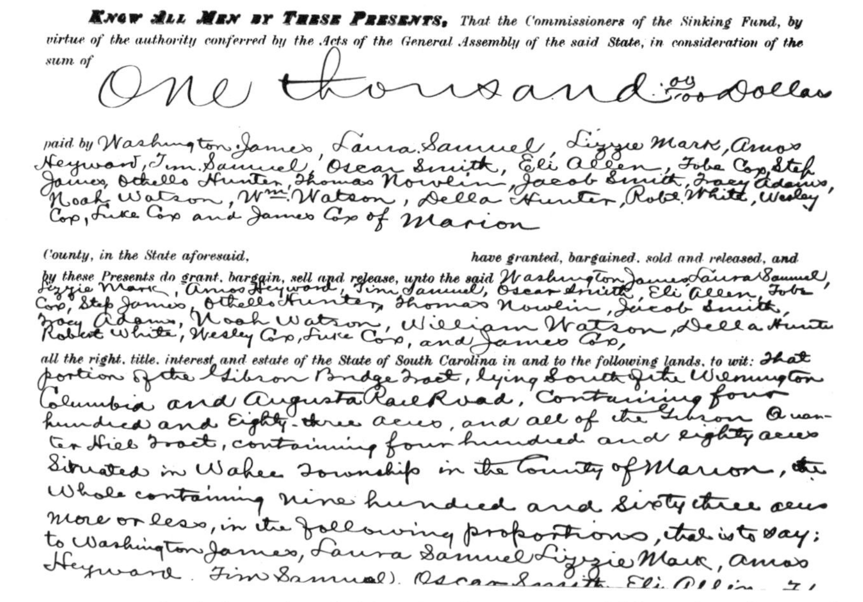

424

The State of South Carolina:

KNOW ALL MEN BY THESE PRESENTS, That the Commissioners of the Sinking Fund, by virtue of the authority conferred by the Acts of the General Assembly of the said State, in consideration of the sum of One thousand 00/100 Dollars

paid by Washington James, Laura Samuel, Lizzie Mark, Amos Heyward, Tim Samuel, Oscar Smith, Eli Allen, Tobe Cox, Step James, Othello Hunter, Thomas Nowlin, Jacob Smith, Tracy Adams, Noah Watson, Wm. Watson, Della Hunter, Robt. White, Wesley Cox, Luke Cox and James Cox of Marion

County, in the State aforesaid, have granted, bargained, sold and released, and by these Presents do grant, bargain, sell and release, unto the said Washington James, Laura Samuel, Lizzie Mark, Amos Heyward, Tim Samuel, Oscar Smith, Eli Allen, Tobe Cox, Step James, Othello Hunter, Thomas Nowlin, Jacob Smith, Tracy Adams, Noah Watson, William Watson, Della Hunter, Robert White, Wesley Cox, Luke Cox, and James Cox,

all the right, title, interest and estate of the State of South Carolina in and to the following lands, to wit: That portion of the Gibson Bridge Tract, lying South of the Wilmington Columbia and Augusta Rail Road, containing four hundred and Eighty-three acres, and all of the Gibson Quarter Hill Tract, containing four hundred and eighty acres situated in Wahee Township in the County of Marion, the Whole containing nine hundred and Sixty three acres more or less, in the following proportions, that is to say: to Washington James, Laura Samuel Lizzie Mark, Amos Heyward, Tim Samuel, Oscar Smith, Eli Allen, ...

Fig. 39 1884 deed from the Sinking Fund to twenty African Americans of Marion County for land just across the Great Pee Dee River from Mars Bluff.

Copied from Records of the Budget and Control Board; Sinking Fund Commission, Public Land Division, Duplicate Titles, B, p. 424, South Carolina Department of Archives and History, Columbia, S.C.

were unable to leave no matter how harsh the terms of employment were. Consequently, European Americans were united in their determination not to sell land to African Americans. James McPherson stated the situation well: "Whites often refused to sell or even to rent them [African Americans] land for fear of losing a source of cheap labor or of encouraging notions of independence."[35]

Sidney Andrews, who traveled through eastern South Carolina just after emancipation, reported that "planters were quietly holding meetings at which they pass resolutions not to sell land to negroes, and not to hire negroes unless they can show a 'consent paper' from their former owner. In Beaufort District they not only refuse to sell land to negroes, but also refuse to rent it to them; and many black men have been told that they would be shot if they leased land and undertook to work for themselves."[36]

Being obliged to live on land owned by European American planters, African Americans were controlled as surely as if they were still slaves. If they did anything to displease the planter, he could say to them: "Leave my land and move your family out of my house." The family would be thrown into the road—with no shelter, no food, no money, no job, and no hope of obtaining any of those things.[37]

European Americans also exerted an opposite kind of control: because the African Americans had no land, they could be held against their will. Many years after emancipation that was still happening. Archie Waiters described an incident from the twentieth century involving the sharecropper Frank Fleming: "One time Frank Fleming and . . . [the landowner] had fall out about something. And he [Fleming] said, 'I'll move.' . . . [The landowner] say, 'You ain't getting a house. Where you going? You about stay here with me.' . . . Told him he couldn't even find no house. . . . And he didn't get no house."[38]

The landowner only had to say the word, and no one had allowed Fleming to have a place to live even though he was known as a man of exceptionally fine character.[39] At Mars Bluff as recently as the early twentieth century, the European American community was still united in its determination to keep African Americans under its economic control.

With that attitude prevailing, it was certain that landowners were not going to let African Americans own land—their best hope for freedom. Consequently, in 1920 Tom Brown faced the same problem he had faced in 1865. Du Bois expressed it this way: "If the basic problem of Reconstruction in the South was economic, then the kernel of the economic situation was land. . . . The main question to which the Negroes returned again and again was the problem of owning land."[40] Surely Brown must have returned to the land question again and again through all those decades of disappointed hope.

Archie Waiters told of conversations he had had with his grandfather about the promise of land. Gregg's hopes were probably similar to Brown's, for the two men were about the same age and had lived in the same community all their lives.

When I first asked Waiters to tell me stories that his grandfather had told him about early times, I expected to hear about things his grandfather had done. I was surprised that the very first thing Waiters mentioned was land. It must have had a high priority in Gregg's thinking. I asked:

> *Q.* What did your grandfather look like?
> *A.* He was a short, heavy-built man.
> *Q.* Would he tell you stories about early times?
> *A.* He'd sit down and talk about slavery times. [Waiters' voice changed to imitate his grandfather speaking.] "You know us—ten acres land and mule. That what us promised."
>
> [Waiters returned to own voice.] Twenty-five cent an acre for land in his lifetime. When they fight this war to free Negroes, promised ten acres land and a mule. Where you get twenty-five cent an acre to buy land with? It take all that for you to live.[41]

Waiters recalled another conversation about the promise of forty acres and a mule. This conversation must have taken place as recently as the 1930s, for in it Waiters talked back to his grandfather—something he would never have done as a child.

> Pa say when they come over here they tell them they cleaning up this land and they get everything settled down they give them a mule and forty acres of land. That what he tell me.
>
> I say, "Pa, they don't have to—Man, where your land at?"
>
> He say, "They ain't give it to me."
>
> I say, "Well you ain't going to get none then. You going to have to be—You ain't going to get none then."
>
> He say, "That what they promised, a mule and forty acres of land."[42]

It would appear that when Gregg was a very old man, time had not dulled the hope and disappointment he felt about the promise of land. Men like Alex Gregg and Tom Brown would have had good reason to trust that promise, for it had been made by both the federal and the state governments.

But neither government made good on its pledge, and European Americans would not sell land to African Americans. As Eric Foner wrote, "The vast majority of blacks emerged from slavery lacking the ability to purchase land even at the depressed prices of early Reconstruction, and confronting a white community united in the refusal to advance credit or sell them property."[43] Consequently the acquisition of land was an insoluble problem for African Americans.

All of the facts surely pointed to the conclusion that it was an insoluble problem. But then I heard the excitement in Rev. Saunders' voice as he described the clearing of land for a rice field. He said, "They made land!" That was what Rev. Saunders was excited about—not that African Americans had made a rice field but that they had made something much more important.

They had chosen a low place in a forest for a rice field. It was land that was worthless to the European American owner because there was no market for the hardwoods that grew there and because European Americans did not raise any crops that would grow on such wet land. African Americans selected these tiny wet spots in the forest and then, by a tremendous investment of labor, turned them into the thing they needed most.

When they had accomplished the herculean task of clearing the land, they must have felt that this half acre or so belonged to them by right of their investment in it.[44] They must have felt that they were working out the second step of their emancipation.

They had secured the land in an African way—by clearing it of forest. In Africa, their ancestors had cleared land when they moved into new country or when their small fields became infertile.[45] Land clearing was also used in Africa by people who wanted to move into a village; they used it as a way of establishing their membership in the community. Melissa Leach, who recently studied a Mende village in Sierra Leone, wrote: "It is land clearance rather than planting and harvesting which is regarded as establishing 'ownership' of land; 'releasing it from the bush,' as it were. . . . Today, an important part of the incorporation of 'strangers' into Mende communities involves their making farms (clearing land) within their stranger-father's kin group area, and this act thus becomes an expression of their 'belonging' to that group."[46]

Similar African concepts probably were a part of Tom Brown's heritage. Paul Richards saw in Brown's land clearing on the day before he died an act that resonated with an African world view. Richards said:

> Brown's work was a way of expressing his manhood, his confidence as an elder, and his responsibility to create a habitable place for his people. It was not landowning in the European

> sense. It had more to do with the possibility of community continuity, for Brown's land clearing was expressive of membership in a community that would naturally think in terms of owning land collectively. In West Africa land is conceived of as belonging to the dead, the living, and the yet unborn. Brown's act expressed the relationship between himself as a member of the wider community and the land which would sustain that wider community.[47]

Since the day of Brown's emancipation, he and all of his people had yearned for land for their own. Brown had been denied land that he had been promised by the government and denied a chance to purchase land from European Americans. Still, he had not accepted defeat. He had come out of a strong, resilient African culture in which every member of the community shared the right to clear and use a small piece of land.[48] Through a difficult period of history, Brown had used that African tradition to make real the hope of his people. On the day before he died, when Brown worked with his neighbors to clear land, he must have felt a deep sense of unity with his people and their aspirations.[49]

Epilogue

WHILE WRITING THIS book, I learned two things—something about who I was and something about interviewing.

What I learned about myself came as a shock. Searching for information about rice growers in Africa, I came upon a book written in 1818. Its author, who was French, summed up what Africans knew about agriculture in one sentence: "All they [Africans] know of agriculture, they have learnt from the Europeans."[1] I was indignant with the author—what a Eurocentric thing to say.

Then one day, years later, a light suddenly went on in my head; I saw that I had made the same error as the Frenchman. I had not asked Archie Waiters about rice cultivation at Mars Bluff, though I had asked lots of European Americans. Because they had not known, I had assumed that no African American would know. At a deep unconscious level, I must have believed that Africans knew nothing that they had not learned from Europeans. I was as bad as the Frenchman that I had been condemning.

My hope is that if I interview enough African Americans and learn more and more of what they have not learned from European Americans, eventually my prejudiced attitude will have to vanish.[2] But I am aware that I wrote this book while harboring deeply ingrained prejudices, and I look forward to the day when the Mars Bluff African American story can be told with clearer vision.

While writing this book I also learned something about interviewing. I began the project knowing nothing about interviewing and nothing about Africa. I could not even recognize Malinke as an African ethnic group. The people I interviewed deserved to have a trained interviewer knowledgeable about Africa and African American culture. But I felt that as long as professional interviewers were not available, the privilege of recording the older people's memories was the responsibility of anyone who was concerned about the loss of the African American cultural heritage.

As an untrained interviewer, I am an authority on the mistakes that it is possible to make. One that I frequently make is initiating an interview with a preconceived idea about what I am going to learn. I have found that people rarely know what I expect them to know. When I persist in asking questions about what I think I am going to hear, both the person I am interviewing and I are frustrated. On the other hand, if I let the subject tell me what he or she knows, I don't get the information I expected; I get information that is far superior.[3]

I am learning slowly that interviewing a person is not like going to an encyclopedia and looking for a certain subject on a certain page. Older African Americans are living treasuries of their own unique African American experiences. When they are generous enough to share their memories with me, I must let them take me in words where they have walked in fact. It is a journey for which I own no road map; in fact, I do not even speak the language of the country we are visiting. Only the person being interviewed can say where we are going—and where we have been.

These elderly African Americans show me glimpses of African American practices and ancient African cultures that I never knew existed. I count it a great privilege when they take me with them in their reminiscences.

Appendixes

Appendix A
African American Words and Customs

FIFTY-FIVE YEARS AGO when I lived at Mars Bluff, many African Americans used words and maintained customs that seemed to have African roots. Perhaps some of those cultural clues can be used to establish connections between Mars Bluff African Americans and specific parts of Africa.[1] The search for connections is exciting because before now so few links between Mars Bluff and Africa have been sought or found.

As I am not an African scholar, I make no claim that the clues I have gathered are scholarly—or even that they point in the right direction. They are simply the findings of one person who believes that the search is worthwhile. It is a reaching out to Africa for a vigorous cultural heritage that has been too long denied. Below are some Mars Bluff words and customs I have explored.

Baby

Baby, yearling baby, and *knee baby* were terms used at Mars Bluff to distinguish which baby a mother was talking about. An infant was called "the baby"; a one-year-old was called "the yearling baby"; the next oldest baby, if a toddler, was called "the knee baby." That three-way classification may not have been of African origin; however, at Mars Bluff, the fact that it was used routinely by African Americans indicates the possibility of an African connection.

Joko Sengova, an African linguist from Sierra Leone, told of a three-way classification of children among the Mende. The term for the youngest child is *ndolga nege-nengei,* a phrase that means "very tender, like a new plant." The toddler is called *jialoi,* a word derived from the words for *walk* and *child.* The third stage is called *ndoi,* a term of endearment for any child up to seven years.[2]

Baby Passed over Grave

Asked about the custom of passing a baby over a grave, Catherine Waiters replied, "They say that to keep the baby from being scary or something like that. I've seen that not long ago at Mt. Zion—after they fill the grave. They say the baby wouldn't cry as much. They still do that."[3]

A recent study of African American burial customs in South Carolina explains: "The purposes for passing a child over a coffin remain unchanged from what they were during slavery—to keep the child from 'frettin,' or being afraid of the dead and to keep the spirit from claiming the life of the child. As a result of taking the appropriate measures, any attempts by the spirit to haunt or harm the child are thwarted."[4]

Biddy

At Mars Bluff, chicks were called "biddies," and the house where they were kept was called the "biddy house." One of the African Americans listed on an 1839 estate inventory at Mars Bluff was named Biddy.[5]

The English word *biddy* comes from the Bantu word *bidibidi,* meaning "small, yellow bird."[6] *Webster's* dictionary fails to mention the African origin of the word, saying that the word is "perhaps imitative."[7]

Booger

Children at Mars Bluff have been told that "the booger man would get them." *Booger,* meaning "ghost" or "spirit," may come from the Bantu word *mbuka,* meaning "consultation of the spirits."[8]

Bottle Tree

When Catherine Waiters was asked if she had ever seen a bottle tree, she said, "On the other side of Timmonsville, I had seen some—a whole lot of bottles hang up in a tree. I was wanting to know what they was doing there. They had a lot of bottles hanging on it. We were going to a singing in a place called Lamar. [The bottle tree was] between Timmonsville and Lamar, on the right . . . a long time ago."[9]

Robert Thompson wrote: "There is a persistent Kongo-derived tradition of *bottle trees*—trees garlanded with bottles, vessels, and other objects for protecting the household through the invocations of the dead."[10]

Cooter

At Mars Bluff, turtles were called "cooters." Leon Coker, describing the skill of the Jamestown hunters, said, "They had men down there could walk through that swamp . . . and they could track a cooter. Reach down in the mud and get it. I mean a big cooter." Words similar to *cooter* are found in several West African languages and in one Bantu language of the Congo-Angola region.[11]

De Monkey

At Mars Bluff, *de monkey* was to collapse from working in excessive heat. The term was used as a verb—"He monkeyed"—or as a noun—"De monkey slap him." It may be of Bantu origin—related to the African word *dipungi,* meaning "exhaustion" or "fatigue."[12]

When Archie Waiters was asked if he ever saw anyone monkey when he was working at Mars Bluff, he said, "Nobody monkey in them days. Nathaniel Smalls working on the railroad, only one I seen."[13]

Dipper

Gourds were used as dippers at Mars Bluff; they were frequently used for drinking water. In Africa, the use of gourds is still common.[14]

Geech

Talking about Tom Brown, Waiters said, "Better not point your finger at him like that; that man would jump through the roof. He's geechy [ticklish]." *Geech,* meaning "to tickle the ribs," may be derived from the Bantu word *kuitsha,* which refers to paddling or poking the ribs with a finger paddle.[15]

Grave Dirt

Rev. Frank Saunders recalled that his boyhood home at Mars Bluff was in a field beside a forest where there was an old cemetery. Frequently, African American men would pass his home. Carrying a sack and a spade, they would enter the woods. Saunders said that he did not know what they were doing; however, their activity seemed noteworthy to him, for he recalled it sixty years later. It is possible that they were going to get grave dirt. "In Kongo territory, earth from a grave is considered at one with the spirit of the buried person."[16]

Graves

Asked what she recalled seeing on graves at Mars Bluff, Catherine Waiters said, "Anything that was of value to the person. Like if it was a vase or a dish or anything like that, they'd put it on." Thompson wrote, "Both Kongo and Kongo-American tombs are frequently covered with the last objects touched or used by the deceased."[17]

Great Da

The origin of the word *Da* is discussed in Chapter 1.

Guinea

A guinea is a noisy fowl that was domesticated in Africa and brought to the United States.[18] Guineas were present on many Mars Bluff farms well into the twentieth century. They had a loud, harsh call that sounded like "pot rack, pot rack, pot rack."

Ida Zanders told about her mother's life after she moved from Mars Bluff to Florence: "My momma had everything right there that she had in the country. . . . She had a cow. She had chicken and ducks and guineas."[19]

Gumbo

At Mars Bluff, gumbo consists chiefly of okra, tomatoes, onion, and a little salt pork. It is eaten on cornbread or rice. *Gumbo,* meaning "okra," derives from Bantu languages.[20]

Hidden Africanisms

Margaret Creel wrote of the Gullah's use of Africanisms in religion: "I do not argue for 'survivals,' a somewhat lifeless term implying passive existence. But I do argue for the presence of dynamic, creative, cultural trends of African provenance."[21] In a similar way, perhaps many Africanisms, both religious and secular, live on at Mars Bluff, merged in the European American culture as African Americans have adapted their African cultural heritage to New World conditions.

Hoppin John

Hoppin John is rice, cowpeas (called "field peas" at Mars Bluff), and salt pork cooked together.[22] Everyone at Mars Bluff—African Americans and European Americans—had to eat hog jowl and hoppin John for dinner on New Year's Day. Whoever did not eat both would have bad luck all year.

Fig. 40 Catherine Waiters with kerchief.

The cowpea is indigenous to Africa. Its cultivation was developed in Africa, and it is still widely eaten there.[23] At Mars Bluff, cowpeas were routinely planted in corn fields beside the rows of corn—a practice probably based on the African custom of intercropping, planting a variety of crops together.[24]

Kerchiefs

Three photographs give a glimpse of how kerchiefs were worn by older African Americans at Mars Bluff. Catherine Waiters shows how her

Fig. 41 Julia Johnson with kerchief.

great-grandmother, Irene Charles, wore her kerchief. Frances Johnson tied her kerchief as her mother had. Julia Johnson did not have to tie on a kerchief for her photograph; she wore one routinely.

Herskovits wrote about the variety of ways headkerchiefs are tied in Africa. He points out that, despite the prevalence of the custom, in the United States, wearing kerchiefs is an Africanism about which we have little knowledge.[25]

The terms *kerchief* and *headkerchief* were not used at Mars Bluff. Instead, a woman was apt to speak of her "head rag."

Fig. 42 Frances Johnson with kerchief.
Photograph © 1987 Sidney Glass.

Lee

Archie Waiters used the word *lee* to describe something very small. If an object was small, he said it was little; if it was very small, he said it was "lee little." For example, when he was describing what he played with as a child, he said he had a "lee little wagon rim and a piece of wire bent to fit over that rim. . . . Push it and run and try to catch it."[26]

Turner wrote that the word *li* is used in Gullah to mean "young," "small," or "recently born." A similar word for a recently born child is found in the Wolof language in Senegal and Gambia.[27]

Look at Her Straight

Matthew Williamson recalled an incident from his childhood that illustrates a practice that seems to have come from Africa. Williamson's teacher had whipped him—unjustly, he thought. He told his mother that he was going to get even with the teacher. "My mother told me, say, 'If you just look at her straight' she [his mother] was going to skin me."[28]

Children were not allowed to look directly into the face of an adult. Was that custom universal in Africa or limited to certain regions? According to Sengova, the custom is widespread throughout West Africa. The African historian Alpha Bah said that even adults are prohibited from looking other adults in the eye because it implies a lack of respect: "For many Africans, looking somebody straight in the eye is an insult. It's telling you that we are equal."[29]

Malinka

The origin of the name *Malinka* is discussed in Chapter 1.

Maum

"Maum" was a title of respect given to elderly African American women at Mars Bluff. Alex Gregg's second wife, a widely respected midwife, was called Maum Florence. The word may be derived from the Bantu *mau* and *mama,* meaning "mother" or "woman."[30]

NAMES OF FIVE ELLISON CHILDREN

Five of the children of Fanny Jolly and Hillard Ellison had names similar to African names. As a child, Mrs. Ellison had lived in Wilmington, North Carolina. There is no way to know if the choice of her children's names was influenced by her Wilmington years or by her Mars Bluff years.

One of the Ellison children was named Cooter. That name could have come from the word for *tortoise* (see *cooter* above). Another possibility is that it came from *kuda,* the word for Wednesday used by the Ewe people in Togo and Dahomey.[31] It is a common practice among many African peoples to name a child for the day of birth. The two words would have sounded alike at Mars Bluff, where *r* is never pronounced and the last syllable of a word is generally slurred.

One of the Ellisons' sons was named Bucka. The word *buckra* comes from *mbakara,* a word for "master" in southern Nigeria. In the United States, *buckra* means "white man."[32]

A daughter of the Ellisons was named Dawra. Turner gave Dara as a girl's name used by the Yoruba people of southern Nigeria. It means "beautiful," "to perform feats," or "habitual." Another daughter was named Twailee. Turner gave Twala as a girl's name used by the Kongo people of Angola. It means "to introduce or to inflict punishment."[33]

The Ellisons also had a son named Shurley, pronounced "Shule." Shule was the name borne by one of the men aboard the *Amistad,* a ship commandeered by slaves in 1839. He was a Mende from Sierra Leone. Paul Richards thought it possible that Shule could be a shortened form of Suleiman, a name that is common among Muslims and therefore not limited to a specific part of Africa.[34]

NAMES OF OTHER PEOPLE AT MARS BLUFF

There is no claim that the names below are of African origin. They are included here because they are strange to the English ear. Probably many of them originated in the United States; however, some of them may one day be traced to African sources. The use of nicknames, especially animal nicknames, was such a common occurrence in earlier days at Mars Bluff that it seems likely to have had an African source.

Bood, a child of Fanny Jolly and Hillard Ellison
Brot, Archie James of Jamestown
Bug, a James of Jamestown
Cuar, listed on an 1802 estate appraisal
Desby, a James of Jamestown
Dort, Chloe Jenkins' sister at Mars Bluff
Duca, Lena Mae Garland
Dutsy, Narcissa [Ford?]), of Jamestown
Epshen, a Smalls, Sara Howard's first husband
Gadney, a child of Fanny Jolly and Hillard Ellison
Lunda, a child of Emma and Alex Gregg
Micie, a James of Jamestown
Minda, Frances Johnson's grandmother
Monk, a child of Fanny Jolly and Hillard Ellison
Nankie, a son of Nellie and Alfred Robinson
Nun, a son of Nellie and Alfred Robinson
Pawthinger, a sister of Fanny Jolly Ellison
Rat, a James of Jamestown
Ressie, a James of Jamestown
Tepee, listed on an 1839 estate appraisal[35]
Tig, Julia Johnson, Frances Johnson's sister
Toutapeg, a James of Jamestown
Vico, Frances Johnson's father[36]

Nini

Catherine Waiters said that her great-grandmother used the word *nini,* and at Mars Bluff it was common to hear a nursing mother offer her breast to her baby saying, "You want to suck nini?" *Nyini* refers to the female breast in the Mende language in Sierra Leone.[37]

Payday

Archie Waiters said that his grandfather's sons worked for the landowner and that every Saturday night their mother would go to the store to get their pay. If her husband was due pay, she would get that too. Waiters said: "Then they didn't pay off until Saturday night now. . . . About eight o'clock Saturday night. . . . Just like you my mother, you go over there [the store] and get that pay. . . . All [would] work. . . . She'd [Emma Gregg] go over there and get the money. . . . She'd get it for all of them. Anybody that worked for him, she'd get it."[38] It is possible that this habit represented a matriarchal orientation that had come from Africa.

Pinda

When Catherine Waiters was asked about peanuts, she said, "Way back a long time ago they used to call them pindas—the older people. I would call them pindas until I get up larger."[39]

On the other hand, Mrs. Waiters did not know the meaning of the word *goober.* Both it and *pinda* are of African origin. The preference for *pinda* seems to be fairly common at Mars Bluff. In that context it is interesting to note that Guy Johnson wrote, "*Goober* (peanut) is common throughout the South, but its synonym, *pinda* or *pindar,* is less popular."[40] The Mars Bluff people's preference for the less common word may say something about their African connections.

African linguist Hazel Carter said that the word *pinda* originated with the Kongo people who lived south of the Congo River in present-day Angola and Zaire. *Goober* comes from a word used by the Kongo people and also by the Mbundu, an ethnic group that lived south of the Kongo.[41] To know that *pindar,* the word used at Mars Bluff, is strictly a Kongo word helps establish a connection between the people of Mars Bluff and the Kongo people.[42]

Swallow Your Tongue

Waiters told about a tomato soup his grandfather had enjoyed eating when he was a slave. "Pa says that thing was good. Put a little bit of sugar in it, a little bit of pepper, a little bit of salt. . . . Pa say that thing would be good. . . . Pa say that thing make you swallow your tongue."[43]

The expression "*swallow your tongue*" is strange to European American ears. St. Clair Drake mentioned tongue swallowing as a method of resistance among African American slaves, but he did not give an ethnic connection. Ulrich B. Phillips wrote that it was a method of suicide used by the Ibo people.[44] Sengova said that the expression "swallow your tongue" was used by the Mende to describe a part of the process of dying, rather than the cause of death. He went on to say that among the Mende, the expression that describes something that tastes very good is, "It is so good that you bite your tongue."[45]

Swept Yards

The yards of African Americans at Mars Bluff were swept daily with a brush broom. They might be swept even more often, for the children were given the task, and when the children played in the yard, there was frequently some small child practicing sweeping in imitation of the larger children. The brooms were made by tying together four-foot branches from a small tree, preferably a dogwood. Usually they were tied together with strips of old cloth.[46] The daily sweeping plus hard use and a high sand content in the soil caused Mars Bluff yards to look as if they were made of concrete.

No one had ever been able to tell me the reason for the custom of sweeping the yard. When I asked African Americans why they wanted the yards swept everyday, their usual reply was, "I can't stand a dirty yard."[47]

Although explanations have been impossible to find in South Carolina, I have gathered information on the West African custom. Perhaps something similar to the West African rationale is the basis for the Mars Bluff custom.

Sengova explained that when he was a child in Sierra Leone, his family had two reasons for sweeping their yard. One reason was to get rid of dirt that accumulated during the day. That had to be done before dark, for there is a taboo against throwing dirt away after dark. However, the important time to sweep was in the early morning, and then the rationale was entirely different.

Fig. 43 Catherine Waiters made this brush broom of dogwood to demonstrate how she used to make them. This one is tied with wire rather than strips of old cloth. Mrs. Waiters no longer sweeps her yard. This clean-swept yard belongs to Cora Robinson, a neighbor of Mrs. Waiters and one of the few people at Mars Bluff who still sweep their yards.

Sengova said, "In the morning as soon as you wake up, the first thing you do is sweep the yard. The witches have had their dances outside of the compound. You have to sweep in the morning so that humans can start to dance their dance of the day. . . . There are nice spirits as well as bad spirits. That is why to be on the safe side you use caution. So first thing in the morning, you have to sweep away all those footsteps and all the remains and remnants that have been put there overnight."[48]

Anthropologist Joe Opala confirmed Sengova's explanation of why yards were swept. Opala had learned the hard way. Years earlier, when he had first gone to Sierra Leone as a Peace Corps volunteer, he had let grass grow around his house. One day a spokesman for the chief came and very sternly told him that he could not allow grass to grow in his yard.

Opala recalled, "They said it in such a way—like 'Only a fool would do that. What's wrong with you?' It is so fundamental to them." Now, after many years in Sierra Leone, awaking to the sound of sweeping every morning, Opala explains the deep cosmological importance of the swept yard.

> By keeping the yard free of vegetation, you are maintaining the cosmic distinction between town and bush. You are maintaining the distinction between an area of order, harmony, law, and ancestral rule and an area of wilderness chaos. . . .
>
> The town is a place . . . where chiefs rule with wisdom and authority, drawing on the wisdom of the ancestors. It is also the place where the ancestors dwell—right there, guiding the people. The forest is a place of disorder, occupied by invisible spirits that are dangerous, frightening, and potentially destructive to the town. It is a world of constant tension between the town . . . and the bush that surrounds every town.[49]

That was Opala's explanation for yard sweeping in Sierra Leone. Alpha Bah, who had lived in Sierra Leone but whose parents were Fulbe people from northern Guinea, offered the Fulbe's perspective: "Everybody sweeps. That's given. Even if it's clean. That's the first job of the morning for the young child, particularly young girls."

When asked the reason for sweeping, Bah replied, "The first thing you do in the morning is to get ready for guests . . . and it's looked upon as bad luck not to clean your yard. It has to be swept every day." Bah explained that it was bad luck for anyone to go visiting so early in the morning that the host had "not brushed his teeth, washed his face, and swept his yard. Those things go together."

Although he could give no reason for the prohibition against sweeping at night, Bah said, "You don't sweep at night. For some groups, you only sweep at night when somebody is dead in the house. When somebody dies, you go to inform the neighbors and they are going to come in. That is the time you are allowed to sweep. You don't welcome people to visit the dead when the place is dirty."[50]

Bah's statement that, among the Fulbe, yard sweeping is a task assigned to young children contrasts with Opala's statement that women do the early-morning yard sweeping in Sierra Leone. At Mars Bluff, women did some sweeping, but more often the task seems to have been assigned to children.

Syrup

Sorghum was grown by some European Americans at Mars Bluff, but the practice was more common among African Americans—as was the practice of drinking sweetening water, or syrup water.

Matthew Williamson showed me where his grandfather, Tony Howard, had raised ribbon cane and sorghum on the low end of a dryland field. He explained how the older people made syrup:

> They stick the cane between those rollers. . . . It squeezes all the juice out of the cane. . . . It had a long pole, the mule tied on one end, and he would go round and round in a circle. . . . That's what turned the mill. One person would feed the mill. . . . Then had another place set up for the cooking which we call a pan and had sections in there so you could push it [the juice] from one section to another.
>
> [Both Tony Howard and Sherman Williamson had syrup mills.] They would grind for other people that would plant cane and didn't have a mill. They would charge for grinding it or sometimes they would pay a toll . . . so much syrup for grinding it.
>
> My daddy [Sherman Williamson] used to make what he

> called sweetening water and they said it was good for you. It would cool your blood down in the summer time.[51]

See Chapter 4, where Archie Waiters tells how to make sweetening water.

"The" Rice

In many interviews, I was struck by the use of the term *the rice* where a European American would have said *rice*. For example, Rev. Saunders said, "I wouldn't try to name [the others], but there was more than them planting the rice."[52]

Perhaps there is an African source for the habit of putting *the* before the word *rice*. Supporting this conjecture is a rice-husking chant formerly used by African Americans in the Georgia Sea Islands. The chant puts *the* before *rice* in the same way that older Mars Bluff African Americans did. The chant begins: "Peas an' the rice, peas an' the rice / Peas an' the rice done done done done."[53]

Yunna

At Mars Bluff, a person would say, "Yunna come in the house," or "That yunna dog?"[54] A *Krio-English Dictionary* gives Igbo as the source of *una* and *yuna*. Both words are the plural form for *you* and *your* in Krio, the English-oriented creole language of Sierra Leone.[55]

Appendix B
Threshing and Husking Rice

AFRICAN AMERICANS AT Mars Bluff threshed rice in a number of ways. Some beat the sheaves with sticks; some pulled the rice off by hand; some put a sheaf of rice in a burlap bag and beat it with a maul.[1]

Threshing Methods at Mars Bluff

Coker, Preston	Put sheaf in a burlap bag and beat it with a maul.
Daniels, Arthur	Wrung the heads off, then beat them with a maul on the floor of the barn.
Ellison, Fanny	Put sheaf in a burlap bag and beat it with a maul.
Gregg, Horace	Pulled rice grains off by hand.
Johnson, Vico	Placed sheaves horizontally on a tablelike scaffold. Beat them with a special floppy flail.
Scott, Willie	Hung sheaves across a horizontal pole, heads down, for about a week. Rice dried and fell off spontaneously. Beat with a flail (any stick) while hanging. Final step: held sheaf in hand and beat it across a horizontal pole three feet above the ground.
Smalls, Lewis	Pulled rice grains off by hand.

The wide variety of threshing methods used by Mars Bluff rice growers compared well with Paul Richards' observation about African farmers. He wrote that present-day Sierra Leone rice farmers "are prepared to probe and attack agricultural problems with a marked sense of inventiveness and flair for experimentation." Richards suggested that the habit of experimentation and innovation might have been a more important contribution to South Carolina agriculture than any specific technique Africans brought with them unchanged from Africa.[2]

Vico Johnson was most innovative; in one step, he accomplished both flailing and some of the winnowing. On the irrigated plantations in the tidewater region of South Carolina, the rice was first flailed and then winnowed in a winnowing house. Duncan Heyward described the method:

> To thresh the rice, the planters had, for many years, to resort to flail-sticks. The bundles of rice were placed in rows on the ground, with the heads joining each other. The Negroes walked between the rows, swinging the flail-sticks above their heads, and bringing them down on the heads of rice, thus beating off the grain.
>
> The use of the flail-stick was slow work, especially as it could be carried on only during good weather. When the grain was threshed off, it had to be gathered up and carried to what was known as the "winnowing house," a building about twenty feet off the ground, supported on posts. In the floor of the winnowing house, a grating was placed through which the rice was dropped to the ground, so that, if any breeze were stirring, it would blow away the light and unfilled grains and also any short pieces of straw which were mixed with the rice.[3]

This two-step procedure was what Johnson's method was designed to accomplish in one step. He flailed his rice on a tablelike scaffold. As the grains of rice dropped to a sheet spread on the ground, some of the chaff would blow away.

Johnson used another ingenious device, a floppy flail that he made by putting the center of a hickory sapling in hot coals. Frances Johnson was adamant in maintaining that when her father flailed rice, he laid the rice sheaves on the scaffolding and hit them with the flail. She rejected the idea that he might have held a sheaf in his hands and whipped it

Fig. 44 This rice-flailing scaffold was constructed in an attempt to conform to Frances Johnson's description of her father's scaffolds. A sheet is on the ground to catch the rice. Miss Johnson saw this photograph and said that the scaffold is too high.

Photograph © 1987 Sidney Glass.

against the scaffolding. A present-day Nigerian drawing shows a table-like scaffold being used for threshing rice. But in the picture a woman holds a sheaf and beats it against the scaffold.[4]

Like Johnson, perhaps Willie Scott intended to accomplish some winnowing while he was flailing. Archie Waiters recalled that when he helped Scott with his rice crop, they would drape the sheaves over a horizontal pole so that the rice hung like clothes on a clothesline. The sheaves hung for about a week, so that some rice dried and fell to a sheet spread on the ground; some was flailed off with sticks. Finally, they removed the sheaves and beat them over a horizontal pole three feet above the ground.[5] In each of these threshing techniques the rice had to fall three feet; so just as in Johnson's method, the lighter chaff blew free as the rice grains fell to the sheet below.

Rice has been threshed in a variety of ways on other continents. S. A. Knapp described Oriental threshing methods at the turn of the century: "The grain is universally cut with a reaping hook, bound in

Fig. 45 A drawing from *Recommended Practices for Swamp Rice Production,* a book used for teaching purposes in Nigeria, shows a woman flailing rice on a tablelike scaffold.

Reproduced by permission of Agricultural Extension and Research Liaison Services, Ahmadu Bello University, P.M.B. 1044, Samaru—Zaria, Nigeria.

bundles about three inches in diameter and hung on bamboo poles or laid on the levees of the fields for curing. The rice grains are then removed by drawing the heads through a hatchell (an instrument with long iron teeth set in a board) or by pounding them over a log, or by piling the sheaves on a clay floor and driving oxen over them, as the custom of the country may approve."[6]

Irvine, writing about West African methods, said: "The seeds are removed from the heads or sheaves, generally by beating them with heavy sticks and later with lighter sticks to remove the remainder. This method is generally adopted when larger quantities are dealt with, although it tends to scatter and waste some of the seed. When only small quantities are to be done, they are separated by treading out the grain with the feet."[7]

What was the source of the Mars Bluff threshing methods? It would seem that African Americans at Mars Bluff devised their own methods. But there is no doubt that as a people they had centuries of experience threshing rice in Africa, and they probably drew on that experience. Perhaps someday more substantive connections can be made between African practices and some of the Mars Bluff threshing methods.[8]

Mortars and Pestles

South Carolina rice growers, both on the irrigated tidewater plantations and on the subsistence plots at Mars Bluff, used wooden mortars and pestles for husking rice. There is ample evidence to support the view that the mortar and pestle are of African origin. Nevertheless, one conflicting claim deserves a hearing, though it involves Louisiana, not South Carolina. Charles Gayarré, writing about the period before the arrival of the French in Louisiana, said that Native Americans cultivated corn, "which they knew how to grind with a wooden apparatus."[9]

Chan Lee gave Native Americans credit for the wooden mortar and pestle used in husking rice grown in *marais* (shallow depressions that held water) in southwestern Louisiana. He said: "Milling of rice was probably the most laborious and tedious job in connection with rice production, and it was done from time to time as family need dictated. It is said that early French farmers of the prairies milled rice only on Saturday afternoons and only for Sunday dinner, because *marais* rice provided enough for the big Sunday dinner, but not sufficient for daily consumption. . . . The wooden mortar and pestle adopted from the Indians were found throughout the Providence-rice culture area."[10]

Native Americans may deserve some credit in Louisiana; however, in South Carolina, the literature in the last decade has spoken with one voice in giving Africa credit for the mortar and pestle used in husking rice. Wood wrote:

> Since some form of the mortar and pestle is familiar to agricultural peoples throughout the world, a variety of possible (and impossible) sources has been suggested for this device. But the most logical origin for this technique is the coast of Africa, for there is a striking resemblance between the traditional West African means of pounding rice and the process used by slaves in South Carolina. Several Negroes, usually women, cleaned the grain a small amount at a time by putting it in a wooden mortar which was hollowed from the upright trunk of a pine or cypress. It was beaten with long wooden pestles which had a sharp edge at one end for removing the husks and a flat tip at the other for whitening the grains.

Joyner made the point that "while the mortar and pestle are universal, they are not universally used in the same way. The lowcountry slaves' manner of pounding the rice was identical to the West African practice."[11]

Littlefield evaluated the possibility that the mortar and pestle used for husking rice in South Carolina originated with Native Americans. He concluded that the English colonists in South Carolina rejected the Indian way of husking wild rice.

> Amerindians did collect the seeds of a river grass called variously wild, or Indian, rice; Indian, water, or wild, oats; or sometimes marsh rye; but this is not a true rice, being classified as *zizania aquatica* rather than as any kind of *sativa,* and it was not cultivated. Moreover, English colonials (in South Carolina anyway) did not adopt the Indian method of husking the grain, though some modification of that process would presumably work as well on true rice; they adopted, rather, the mortar and pestle—in the African way.[12]

Wood, Littlefield, Joyner, Judith Chase, and Dale Rosengarten have all stated that in South Carolina the use of the mortar and pestle for husking rice is of African origin.[13] I believe that they are correct in that attribution. Still, I will not rest altogether easy until I find an explanation for a major discrepancy. African mortars that I have seen pictured, both those from early days and those currently in use, are tapered in a vase shape and have a small pedestal base. Most of the South Carolina mortars that I have seen pictured have straight sides and no pedestal base.

In her book on African American crafts, Judith Chase includes photographs, one from Africa and one from South Carolina. The African picture is captioned, "Ashanti brass weight in the form of a man pounding grain with mortar and pestle similar to those used on the rice plantations of South Carolina." The mortar in the picture has tapered sides and a pedestal base. On the other hand, the mortar from South Carolina pictured by Chase has straight sides and no pedestal base. These characteristic shapes are found in almost every rice mortar I have seen pictured from Africa and South Carolina.[14]

According to Opala, the people of Sierra Leone also comment about the difference in the shape of the mortars. When he shows them a picture of a Gullah beating rice, they regularly say that the straight-sided mortar looks funny to them. Theirs are always shaped like an hourglass. Opala explained, "It's decorative. Nobody would think of not decorating it. The decoration can be very elaborate, like geometric designs around the waist of the mortar. . . . In Sierra Leone, the mortar is the first thing you see when you walk in a compound. They would want it to look nice."

Searching for other possible reasons for the difference in shape between African and South Carolina mortars, Opala suggested that a tapered mortar would be lighter than a straight-sided one. Africans want to be able to move their mortars around easily in the compound, or move them out of the rain, or take them to the field. Lightness would also be an asset when turning the mortar over to pour the rice into a fanner. Opala suggested that Africans who came to South Carolina would have wanted to make vase-shaped mortars but would have been discouraged from doing so because in commercial rice production it was not cost-efficient.[15]

In Bambur, Nigeria, the village where agricultural specialist Iliya Likita lived, all the mortars were vase shaped; however, he said that he had seen straight-sided mortars sitting under big trees where sorghum was grown. He associated them with community use, whereas he associated the vase-shaped mortars with domestic use.[16] Maybe the shape of the mortar depends on something as simple as Likita's explanation, or maybe it depends on local customs in different areas of Africa. Still, the

Fig. 46 When asked how to make a rice mortar, Archie Waiters always began by saying, "Find a hollow gum tree." In this photograph, Otis Waiters is learning why his father said that. Hollowing out a tree is tough going.

marked difference between the shape of mortars in Africa and in South Carolina deserves further investigation. Ever the optimist, I would hope that the search for straight-sided mortars in Africa might lead to clues for Africa-South Carolina connections.

Despite the difference in shapes, there can be little doubt that the concept of the South Carolina mortar and pestle came from Africa. How

Fig. 47 Whereas rice mortars made in South Carolina generally have straight sides, this nineteenth-century spice mortar made by African Americans for Dr. Robert Harllee's home use at Mars Bluff has the vase shape so often seen in African mortars.

else could the mortar and pestle have been made and used with such skill by so many widely scattered African Americans? Surely this was their own method, perfected through generations of use in Africa. As Lydia Parrish has shown, Africans even supplied songs to accompany the blows of the pestle.[17]

Two Mars Bluff African Americans I interviewed followed the prac-

Fig. 48 In spite of hands badly crippled by arthritis, Archie Waiters made this mortar and pestle in the 1980s to show how they were made at Mars Bluff in the 1920s. He said that he made the mortar this small, not because this was the biggest hollow gum tree that he could find, but because this is the best size. A small mortar is easier to lift for pouring the rice. He wanted the mortar about an inch bigger than the pestle so the rice would have room to come up around the pestle.

Photograph © 1987 Sidney Glass.

Fig. 49 Waiters teaching old skills. Three generations gather in Archie Waiters' back yard as he supervises his sons Otis Waiters and Richard Waiters in making a rice maul like the one he made when rice was grown at Mars Bluff. The name of this rice-husking tool is really *pestle,* but that word is rarely heard at Mars Bluff. Front row *left to right:* Rondell Waiters, Otis Waiters, Richard Waiters; back row: Annie Lee Waiters Robinson, James Waiters, Johnny Ellison, Archie Waiters.

tice of putting corn shucks in the mortar. The others who were asked said that they had never heard of using corn shucks.

Frances Johnson was the first person to tell me about the practice. She described how she would husk rice using a mortar and pestle. I thought she was through with her account of husking when she surprised me by adding:

> *A.* Put it [the rice] in a dishpan and fan all that trash out it. Then you got to put it back and you get it good and clear; then you get some shucks and tear the shucks up and put the shucks in the rice.

Fig. 50 This is the rice maul, or pestle, that Otis Waiters and Richard Waiters were making in the previous photograph.

Q. Do what? . . . What are you talking about?
A. [Laugh] . . . Why you want to know something like that?
Q. . . . Tell me about those shucks.
A. . . . Corn shucks. What they shuck off the corn.
Q. What do you do with them?
A. Tear them up, and make them about that big I reckon—with the long string—

Fig. 51 Archie Waiters shows equipment that is used to split wood: a glut, a maul, and an ax. This regular maul and the rice maul are sufficiently similar to explain why, at Mars Bluff, a pestle for pounding rice is called a rice maul. Waiters made the maul; the glut was Ruska Gregg's. The oak tree in the background was one of the trees in Alex Gregg's yard when Waiters was a child.

Q. About three-fourths of an inch across?
A. They be about that wide, I reckon. [She illustrated a half inch with her fingers.] And you tear them like that and throw them in your mortar, in your rice, and you beat that.
Q. Why?
A. To try to clear the rice.
Q. Would they be green shucks or dry shucks?
A. No, dry shucks, off the corn—out the barn.
Q. And it's the fiber in them would work against the rice, or what would they do?
A. I think that's what—to clear the rice. . . . Clear it up. . . . Make it pretty. . . . Make it white. I used to do that more times than a little.[18]

Hester Waiters also reported that she used the corn shuck technique. She thought that putting corn shucks on top of the rice not only helped in making the rice white but also kept the rice from cracking and breaking.[19]

A similar account of the use of corn shucks was given by William Henry Davis, who was seventy-two when he was interviewed in 1937. He lived just across the Great Pee Dee River from Mars Bluff.

> Oh I beat rice many a day. Yes'um, beat rice many a day for my grandmother and my mamma, too. Had a mortar and a pestle that I beat rice with. They take big tree and saw log off and set it up just like a tub. Then they hollow it out in the middle and take (instrument for grinding) pestle that have block on both it end and beat rice in that (tub) mortar. Beat it long time and take it out and fan it and then put it back. The last time it put back, tear off some shucks and put in there to get the red part of the rice out and make it white.[20]

Fanner Baskets

In the South Carolina low country, fanner baskets were closely associated with rice cultivation. A fanner was a flat basket—actually more like a tray with a small lip on it—used for winnowing rice; when the rice was tossed up and fell back on the basket, the wind would blow the chaff away. Making fanner baskets from coiled grass was so much a part of low-country African American life that no one ever questioned what ethnic group had brought the skill to South Carolina. The basket-making tradition was universally accepted as of African origin.

Dale Rosengarten, who made a study of coiled-grass baskets used for winnowing rice, found the evidence beyond question: the practice came from Africa. Even today, near Charleston many African Americans continue the tradition of coiled-grass basket making.[21]

I would expect that in Louisiana the fanners were made by African Americans, for Africans taught the people of Louisiana how to grow rice and African Americans formed a large part of the agricultural labor force. But Lee wrote of the people who grew subsistence rice in southwestern Louisiana, "the rice was winnowed with Indian basketry trays made of cane."[22] Perhaps in some parts of Louisiana, Native American fanners were more available than those made by African Americans.

Mars Bluff African Americans said that they used dishpans as fanners. None of those I spoke with had any knowledge of how to make a fanner or of anyone using a basket for fanning, but these were people who were questioned in the 1970s and 1980s. Perhaps if an earlier generation had been asked, they might have known about fanner baskets.

Appendix C
Soils in Mars Bluff Rice Fields

I OBTAINED DESCRIPTIONS of nineteen of the twenty-two African American rice fields at Mars Bluff. Using that information, I divided the rice growers into two groups, twelve in the wetland group and seven in the dryland. Wetland rice growers are defined by one characteristic: they intend that their fields be submerged, that the fields be under water. On the other hand, dryland rice growers, while their fields may be quite wet, do not intend that their fields be submerged.[1]

In South Carolina, rice cultivation was usually associated with irrigation, so one would have expected to find all twelve of the Mars Bluff wetland rice fields beside streams. Polk Swamp Canal and Middle Branch run through the heart of Mars Bluff and were available to the rice growers. But African Americans did not put their rice fields beside those streams.

Instead, five rice growers chose land at the edge of Polk Swamp that was most distant from the canal. One chose land in a similar location at the edge of the Middle Branch swamp. Three chose land that was altogether removed from the swamps. Three chose land close to Polk Swamp Canal, but Waiters said that they did not use canal water for irrigation.[2]

Although all of the fields were on low ground, most were quite near land with a seasonal high-water table of six feet or more. Why had African Americans put their rice fields in such unlikely spots? Richards wrote of the Mende custom of referring to rice fields by soil type.[3] Perhaps it was soil type, not the availability of water for irrigation, that determined the African Americans' choice for the location of their rice fields.

Almost all of the wetland rice growers chose Pantego soil, which is typically ponded or very poorly drained. There were two exceptions. Alex Gregg chose a perennial pond for his field, and Willie Scott chose Barth soil, which is not as water retentive as Pantego.[4]

Scott's choice seemed strange, but there was a reason for it. His rice field had originally been cleared and used as part of a dryland field by the European American owner of the land. When the owner found that it was too wet to use for dryland crops, he abandoned that part of the field. Then Scott made it into a rice field, saving himself the herculean task of clearing a forested area.[5] During dry periods, however, he was obliged to water his field—while none of the six Pantego fields in that immediate area required watering.

The same explanation holds for Scott's reliance on a small ditch for irrigation in dry periods. He did not choose the site or the ditch because they were perfect for rice; they just happened to be what was readily available. Scott dug two wells in the ditch so that he would have a steady supply of water.

Two other wetland rice growers, Tony Howard and Sherman Williamson, also had to water their fields in dry spells. Both used Pantego soil, but their fields were located above the Middle Branch swamp area, very close to their homes. They had apparently sacrificed the guarantee of a dependable natural water supply in order to have their fields convenient to their homes.

Dryland rice, or upland rice, is classified separately from rice grown in submerged fields. The term *dryland* is misleading; it suggests that the grower wants a dry field, so it seems incongruous that the Mars Bluff dryland growers all sought out wet land. Repeatedly in the interviews, the dryland rice fields were described in phrases like "a wet spot" or "stiff, wet land." Four people spoke of the rice grower using a low, wet place in a dryland field, and the other three fields were also chosen because the soil was wet. Hester Waiters said that Horace Gregg had his field in "a low place. It [rice] grow better in bottoms." Vico Johnson used the lowest land on his farm for rice; his daughter said that "he had it down there in a stiff place." Arthur Daniels' rice field was classified as dryland without my ever seeing it because his daughter likewise described it with the characteristic words for defining a Mars Bluff dryland rice field. She said that it was not covered with water; it was a "dark, stiff, wet piece of land."[6] So dryland, when applied to rice fields, simply means that the land is not flooded or submerged. Rice thrives on water, so even dryland farmers want plenty of moisture.

Soil and Water Conditions in Wetland Rice Fields

Grower	*Soil Type*	*Water Table*[a]	*Watered with Buckets?*
Brown, Tom	Pantego	0–1′	
Fleming, Frank	Pantego	0–1′	
Gregg, Alex	perennial lake or pond		
Harrell, Alex	Pantego	0–1′	
Hayes, Jim	Pantego	0–1′	
Hayes, Willie	Pantego	0–1′	
Howard, Tony and Robert Gregg	Pantego	0–1′	yes
[last name unknown], Jim	Pantego	0–1′	
Robinson, Alfred	Pantego	0–1′	
Saunders, Lawrence	Pantego	0–1′	
Scott, Willie	Barth	$1\frac{1}{2}'-2\frac{1}{2}'$	yes
Williamson, Sherman	Pantego	0–1′	yes

Source: *Soil Survey of Florence,* 11, 18, 78, 82.

[a] Refers to seasonal high-water table.

Soil and Water Conditions in Dryland Rice Fields

Grower	*Soil Type*	*Water Table*[a]
Bailey, Bennie	Goldsboro	$1\frac{1}{2}'-2\frac{1}{2}'$
Coker, Preston	Goldsboro	$1\frac{1}{2}'-2\frac{1}{2}'$
Daniels, Arthur[b]	Pantego	0–1′
Ellison, Fanny	Goldsboro	$1\frac{1}{2}'-2\frac{1}{2}'$
Gregg, Horace	unknown	
Johnson, Vico	Coxville	0–1′
Smalls, Lewis	where Pantego joins Barth	0–1′ $1\frac{1}{2}'-2\frac{1}{2}'$

Source: *Soil Survey of Florence,* 6, 7, 11, 12, 18, 78, 80, 82.

[a] Refers to seasonal high-water table.
[b] Daniels' field may have been a wetland field in wet years. His was the only Pantego soil among the dryland fields.

Fanny Ellison's experience may illustrate the dryland rice grower's ability to adjust to various circumstances. When she lived at Mars Bluff, she planted rice in the lowest part of her vegetable garden, on soil that would have been quite moist for part of the growing season. When she moved to Florence, where her entire garden was on well-drained land, she planted rice in dry soil. Mrs. Ellison's actions seem to say that the dryland rice grower could use dry soil but, if given a choice, would choose a wet spot.[7]

Appendix D
Variety of Rice Planted

THE PEOPLE INTERVIEWED at Mars Bluff could give no name for the variety of rice that had been planted, but Archie Waiters gave one clue. He said that the rice that Willie Scott grew had a blue streak down the side of the grain after it was husked and that the blue streak disappeared when the rice was cooked.[1] Surely such a strange characteristic would be immediately identifiable by anyone familiar with varieties of rice. I expected to learn a name for this variety easily, but in all my years of asking, I have not found anyone who has heard of such a rice.

The odds are that Scott was growing Blue Rose rice, for that was the variety that agricultural bulletins and seed catalogs recommended for subsistence rice growers in the twentieth century. A 1943 Clemson College bulletin encouraging South Carolina farmers to grow upland rice stated, "Most of the rice being grown in South Carolina is the old standard variety, Blue Rose, which is a lowland or irrigated variety but is being grown as an upland variety."[2]

Jenkin Jones, writing about upland rice throughout the South in 1943, spoke of the prevalence of the Blue Rose variety. "The varieties grown as upland rice in the South are, with a few exceptions, the same as those grown with irrigation in the commercial rice areas. Blue Rose is the variety most commonly grown on upland in recent years. Occasional fields of a variety similar to Carolina White are found in South Carolina."[3]

In a final effort to learn what variety of rice had a blue streak that disappeared when it was cooked, I wrote to Nelson Jodon, who had served for many years as a rice geneticist with the Agricultural Research

Service of the United States Department of Agriculture and at the Rice Research Station in Crowley, Louisiana. He replied: "I am unaware of blue color in rice bran. Bran colors include purple, red, brown, and rarely gold besides the brownish color of the bran of ordinary rices which lack those pigments. Rice varieties tend to have a longitudinal groove. If this is pronounced it is difficult to remove the bran from it in milling, thus leaving a streak. It might disappear in cooking if due to red pigment. Red rice has distinct grooves making it very objectionable to the miller."[4] Now I knew one reason why Willie Scott's rice might have had a streak on it—but I still was unable to say definitively what variety he planted.

Paul Richards raised another question: Was there any evidence that the Mars Bluff fields were planted with short-duration rice varieties, that is, early-maturing rice that ripens in about 100 to 120 days? Such crops were an important part of rice cultivation in Sierra Leone. Archie Waiters said that Willie Scott planted rice in March and harvested it in the fall; it was not an early-maturing rice. Although I did not ask the other people who had told me about growing rice, nothing that they had said had indicated that it was early maturing.

In addition to looking for short-duration rice in the United States, Richards was looking for evidence of the African rice *Oryza glaberrima*. He had found evidence in Portères that African varieties of rice had been cultivated in Central America, and he had found intriguing information in the writings of Thomas Jefferson.[5]

In the late 1780s, Jefferson became interested in growing upland rice. He wanted to experiment with rice that could be cultivated in dry fields in the United States, and he asked a number of people to send him seed rice suitable for such cultivation. Nathaniel Cutting sent Jefferson one shipment of seed rice from Africa on July 6, 1790. Cutting wrote: "A Mr. Holman, who for many years has held a Factory for his own account in the river Denby, about the Latt. 9.°30′ North, procured me a ten gallon Keg of that Rice which he calls *the heavy upland Rice.* I have the pleasure to forward it to you herewith."[6]

Richards, who has worked in Sierra Leone and Guinea, says that even today some of the dryland rice grown there is *O. glaberrima*. Defining the Dembia River region as "now in the Republic of Guinea, north of Sierra Leone," Richards wrote about the rice that Jefferson received from that region. "What we don't know was whether the species was *Oryza glaberrima* or *Oryza sativa* since there are numerous dryland cul-

Fig. 52 Varieties of rice. At the top is the African rice *O. glaberrima,* gathered in El Salvador by Dr. Elaine Nowick. It is the only *O. glaberrima* that I can find that was obtained in the New World. At the bottom is the Asian rice *O. sativa,* of the Carolina Gold variety. This sample is courtesy of Dr. Richard Schulze of Turnbridge Plantation, in South Carolina. Dr. Schulze regretted that no one had grown Carolina Gold since 1927, so he began growing it in 1986.

tivars of *both* species of rice on the Guinea/Sierra Leone coast." This raised the question of whether all of the upland rice that Jefferson received from Africa was *Oryza sativa*.[7]

Jefferson was especially interested in dryland rice cultivation in South Carolina because he was concerned about "the pestilential and mortal fevers" associated with wetland rice cultivation. Several of his letters tell of his experiments using seed from Africa to grow upland rice in the United States, especially in South Carolina. On November 27, 1790, he wrote: "About two months ago I was fortunate enough to recieve [*sic*] a cask of mountain rice from the coast of Africa. This has enabled me to engage so many persons in the experiment as to be tolerably sure it will be fairly made by some of them."[8]

Jefferson wrote to William Drayton of South Carolina on May 1, 1791: "I delivered to mr Izard a barrel of Mountain rice of last year's growth, which I received from the island of Bananas on the coast of Africa & which I desired him to share with you for the use of the [agricultural] society."[9]

Seventeen years later, Jefferson wrote a letter telling about the experiments with dryland rice in South Carolina and Georgia: "Nothing came of the trials in South Carolina, but being carried to the upper hilly parts of Georgia, it succeeded there perfectly, has spread over the country, and is now commonly cultivated; still however, for family use chiefly, as they cannot make it for sale in competition with the rice of the swamps. . . . It has got from Georgia into Kentucky, where it is cultivated by many individuals for family use."[10]

At that point in his letter, Jefferson added, "I cultivated it two or three years at Monticello, and had good crops, as did my neighbors, but not having conveniences for husking it, we declined it." Richards noted that present-day West Africans share Jefferson's concern with the problem of husking that type of rice: "In West Africa today the red-skinned country rices are known both for their satisfying flavour and for the inconvenience they cause women in pounding them clean."[11]

The significant point is that dryland rice was brought from Africa to South Carolina and Georgia, and it was grown successfully for a number of years in upcountry Georgia—so I go on searching for evidence that *O. glaberrima* was cultivated in dryland fields in the interior of South Carolina. On this score, Elaine Nowick at the Crowley rice research station sent me a few grains of *O. glaberrima* obtained in El Salvador with a letter saying:

Dr. Charlie Bollich of the Rice Research Station in Texas has forwarded your letter to me for response as I have done some collecting of wild species of rice in S. America and some research on the other wild species of rice.

I do not know if *Oryza glaberrima* was ever grown in S. Carolina but it was collected in El Salvador. . . . I understand that it was brought from Africa by slaves and has since disappeared from this area. I don't believe that *O. glaberrima* would survive long as a wild species because of its adaptation to cultivation.[12]

Appendix E
The Acadians and Subsistence Rice

IN LOUISIANA, WETLAND subsistence rice was called "providence rice." Planted in low spots where water collected naturally, the rice depended entirely on local rainfall for its water supply.[1] That was the same method of wetland cultivation that the Mars Bluff African Americans used; and at Mars Bluff, the practice definitely seemed to have come from Africa. Reading about providence rice, however, one gets the impression that it was not Africans but Acadians who brought this method of cultivation to Louisiana. Is that true?

To get the true story, it is necessary to go back to the early 1700s, when a mere handful of Frenchmen were trying to establish the colony of Louisiana. They were starving. Gayarré told of their struggle to survive: "Thirty-five colonists died of starvation in 1705. . . . In 1709 famine re-appeared in the colony, and the inhabitants were reduced to live on acorns. . . . The scarcity of provisions had become such, that in 1710 Bienville informed his government that he had scattered the greatest part of his men among the Indians, upon whom he had quartered them for food."[2]

Faced with the recurrent problem of starvation, the officials in charge of the Louisiana colony made a rare admission. They acknowledged that they needed African slaves, not just for their labor but also because they knew something that the French settlers did not know—how to grow rice.

In 1718, the Company of the West, which had recently assumed responsibility for the Louisiana colony, gave orders to Sieur Herpin, captain of the *Aurore,* to go to Guinea and other points south of there to trade for four hundred Africans. Herpin's orders read: "In the different places where the said Sieur Herpin trades negroes he will manage to trade for a few who know how to cultivate rice. He will also trade for three or four hogsheads of rice suitable for planting, which he will deliver to the directors of the Company on his arrival in the [Louisiana] colony."[3]

At the same time, the Company of the West gave orders to Sieur Du Coulombier, captain of the *St. Louis,* to go to Angola to trade for 450 or 500 Africans. His orders read: "Sieur Du Coulombier will take measures to trade for a few negroes who know how to cultivate rice. He will also trade for three or four barrels of rice suitable for planting, which he will deliver to the directors of the Company on his arrival in the [Louisiana] colony."[4]

In this way, the desperate colonists sought the help of Africans to save them from starvation; but once out of danger, the colonists made no reference to the Africans—no acknowledgment that it was Africans who had taught them how to cultivate rice.

Those early Africans probably taught the French that one of the easiest ways to make a good crop of rice was to plant a small field in a low place where it would be flooded by rain. That was an African method of cultivation, but it would not have occurred to Africans to call the crop "providence rice." They viewed all of their crops as dependent on the beneficence of providence. All African farming was done in a spirit of mystical relation with the earth and the earth goddess, who supplied what was needed.[5]

To the French colonists, who were in the habit of closely managing their crops, the idea of relying on the providence of God to flood a field with rainwater would have seemed strange. It might well have been those early French colonists who gave the name *providence rice* to this method that the Africans taught them.[6]

Later, Louisianans said that the method was obviously of French origin because when people referred to it, they gave the word *providence* the French pronunciation.[7] The truth probably is that the name is of French origin but the practice is of African origin.

Lee wrote, "On the basis of one of the earliest descriptions of rice culture . . . written possibly in 1725, the earliest form of rice culture in Louisiana seems to have been Providence rice." Reporting on conditions in Louisiana in 1728, Gayarré wrote, "Rice, tobacco, and indigo were cultivated with success by the two thousand six hundred negroes who had been imported."[8] Surely it was those Africans who brought the providence method of rice cultivation to Louisiana.

How, then, did it happen that Acadians received credit for the method? They did not arrive in Louisiana until 1765, almost a half century after Africans had been sent to teach the starving Frenchmen how to grow rice.

The Acadians had lived in Nova Scotia as small farmers for many years before they were deported by the British in 1755. "Most of them were transported to different States. . . . In 1765 about 650 Acadians had arrived in New Orleans." Certainly Acadians had never cultivated rice in Nova Scotia. Although Native Americans in Canada harvested wild rice, wild rice is in fact not rice, nor was it cultivated by Native Americans. In Nova Scotia, the Acadians had raised cattle, hogs, sheep, fodder, wheat, oats, rye, barley, hemp, flax, vegetables, and apples[9]—but not rice.

Before arriving in Louisiana, some of the Acadians had lived for a short while in other colonies.[10] They may have learned the providence rice method there and brought the practice with them when they moved to Louisiana. But even if that was true, the method had already been introduced into Louisiana by Africans a half century earlier.

Quite likely, the Acadians learned the providence method of rice cultivation after their arrival in Louisiana. Known for their music, cooking, and *joie de vivre,* the Acadians also had more serious attributes. They were more egalitarian than many European Americans, and they practiced subsistence farming—taking from the earth only what was necessary for self-sufficiency. For those reasons, they could readily have adopted the providence method from African Americans and incorporated it into their own way of life.[11]

Many Acadians moved into southwestern Louisiana, and they lived there undisturbed for more than one hundred years. Then in 1881, a railroad was built from New Orleans to Houston. It ran through sparsely settled country, "inhabited mostly by Cajuns who fished, trapped, raised livestock and grew subsistence crops." Land developers were hired to promote development of the area served by the new railroad, and large numbers of midwestern wheat farmers were enticed to move to Louisiana. "The immigrants quickly observed that Cajuns grew small plots of rice, so new settlers learned to cultivate rice by the 'Providence' method that relied on rainfall instead of irrigation. . . . For a decade these pioneers relied on Providence to supply irrigation, just as the Cajuns had done."[12]

Then the new settlers introduced irrigation and mechanization. "In the space of ten years the prairies of Louisiana had been transformed from barren land sparsely populated with Cajuns, their cattle, and Providence stands of rice into a booming and highly mechanized rice-growing area."[13] That was how the story was told when the Acadian providence rice growers came into contact with people who were economically motivated to write about crops that could be raised along the route of the railroad. And because so little had been written about providence rice in the past, the land developers were free to write the story as they saw it.

They wrote the story placing great emphasis on the Acadians' role in the growing of rice. Every generation that followed repeated the story—always stressing the Acadians when they wrote of providence rice. Today, the story is still being told in that way.[14]

All those retellings of the story have permanently etched in the literature the idea that providence rice cultivation was primarily an Acadian practice. And it is generally assumed that Acadians deserve credit for introducing the providence method of rice cultivation into Louisiana.

A different story would surely have been told if the land promoters had made a careful survey of all the providence rice growers. They would no doubt have found that as much providence rice was being grown by African Americans as by Acadians. But none of the writers mentioned African Americans.

In 1960, when Lee retold the story, he seemed to foreclose all hope for African American inclusion in the Louisiana providence rice story. He wrote: "In summary, the development of Providence rice has been closely related to the French and their settlements in Louisiana. The known areas of Providence-rice culture are found in French-settled sections, roughly within a triangular area connecting New Orleans, Point Coupee, and Lake Charles. Field investigations indicated that all of the former Providence-rice farmers encountered are of French descent."[15] It was a definitive assignment of credit; "*all* of the former Providence-rice farmers encountered are of French descent." Not one African American.

Is it not more likely that the providence method of rice cultivation was brought to Louisiana by Africans, named by the early French settlers, and later adopted by Acadians? Although the literature credits

Acadians with the establishment of the practice, the evidence actually supports the theory that the providence method of rice cultivation came to Louisiana from Africa.

Whereas South Carolina could correct the telling of its rice story merely by putting Africans into an existing void, Louisiana would have a more difficult task in writing a new version of the providence rice story. First, the Acadian–providence rice connection would need to be loosened; only then would there be a void for the Africans to fill.

Appendix F
Interviews About Rice Cultivation

EXCERPTS FROM FIVE interviews with Mars Bluff African Americans are recorded here in detail. The material is limited to what they had to say about rice cultivation, threshing, and husking. The recording of this material is a small attempt to help halt the obliteration of knowledge about indigenous African-American farming practices. For years those practices have been devalued and left unrecorded. Only recently have agriculturists begun to appreciate the importance of preserving such knowledge. D. Michael Warren and Kristin Cashman, pioneers in this field, have stated the problem well: "A process similar to plant genetic obliteration occurs with the world's store of knowledge whenever local systems of knowledge and technology are suppressed or lost."[1]

I have recorded here interviews with the four subjects who had the most firsthand knowledge of subsistence rice cultivation. For that reason both of the accounts of wetland rice happen to be about fields that required watering. That gives a misleading impression, for out of twelve known wetland rice fields at Mars Bluff only three were known to have required watering. It is also misleading that more dryland fields are described here than wetland fields. Actually, there were only seven known dryland fields while there were twelve known wetland fields.

Mrs. Zanders' interview is included even though she had no direct knowledge of the cultivation procedure. She gives a picture not seen in Mars Bluff's other rice growers. It is a picture of a woman growing enough rice to feed her family simply by planting rice in her dryland vegetable garden.

ARCHIE WAITERS

Lived: 1914–1990
Description: thin, with serious expression, very observant
Interviewed: February 20, 1982, and July 31, 1987, at Mars Bluff
Rice grower: a neighbor, Willie Scott
Method of rice cultivation: wetland
Period of rice cultivation: mid-1920s

[Waiters' interviews have been cut and reorganized extensively to make the material more concise and more comprehensible. Although the order of the material has been changed, the content and wording are true to Waiters' original statements.]

Q. Tell me how you all grew the rice [at Willie Scott's].
A. First get a plow and break the land up . . . somewhere round about the last of January. . . . Then go back about the last of February and lay your land off . . . about a foot deep. . . . And go back and sow your fertilizer: cottonseed meal, kainit, and acid. Just sow it and let it stay there for about three weeks. Then go and sow your rice. . . .
Q. You sow the rice then in March?
A. It be March, the first of March. . . . just sow your rice; don't cover it . . .; and [the rain] runs all the . . . rice all back down in the middle of the row.
Q. How wide apart were the rows?
A. Two-foot rows. . . . Just take and sow it over everything. That big rain come and wash everything right back down. [All the rice grows in the bottom of the furrow even though it was sowed, not planted in rows.] That thing be so thick you couldn't walk between it. . . .
Q. Tell me when you'd start watering it.
A. If it don't rain, you start to watering it. . . . But you didn't have to bother because it rained back in those days; you don't have to start to watering it until it [the water level] start to getting below the bed. Water get about two inches below the bed, then you go there and fill it back up. . . .

Sometime you didn't water it nary a time a month. Sometime watered it twice a month. Sometime nary a time, go on like that. You see, it rained back in them days. . . .
Q. Tell me how you set up those wells.

A. Dig a hole in the ground [in the bottom of the ditch]. . . . Put a chain, two bucket on it, each end of chain. . . . Have a ten-inch board, bottom and top, and have a head back there so the water wouldn't run out, on back end of it. [He is describing a trough made of ten-inch boards on bottom and sides and rear end. The trough carried the water over the little embankment beside the ditch.]. . . .

Have two buckets. I draw a while. You take and pour a while. I get tired of pulling up, you go there and pull up; I pours a while. . . .

Q. And you'd keep pouring until the whole field was up to the top of the beds [in water]?

A. Up to the top of the beds. [When the water would] start to run over the top of the beds, then you cut it off and hang your bucket up and come on back. . . . Time you get done you hear it thundering, RARARARARARA. The water be running out, running back in the ditch. . . .

Q. Then when would you harvest it?

A. Harvest? The fall of the year. About the last fall of the year.

Q. How would you go about that? . . .

A. Had to let the water off. . . . Two or three days [after] you let the water off the rice . . . then the rice will start to laying down. Then you take your hand, you catch a handful, I catch a handful, I'll cut and hand you. Cut another handful and hand you. Cut another handful. You got a bundle about that big now [about fifteen inches across]. Then I cut you about three of them thing [rice stalks] and take and tie it—like that far from the butt end. . . .

That rice taller than you is. [Sheaf is five or six feet tall and is tied about two and a half feet from bottom.]

Q. . . . what did you cut the rice with?

A. A reef hook [a sickle].

Q. What's it look like?

A. It look like a . . . you seen a keyhole saw? It made in a crook like that, a crook like that and got a wooden handle going back here. It got a keen blade, and the further it come back this a way the thicker the blade gets. . . .

Take your hand and hold the rice like this. You get a handful, cut it like this. If you get too much it will slip out your hand. Then you have to stop and pick that up. . . .

Q. And then the rack you go hang it on is up near the house. . . . You all would carry it up by hand?

A. Get the wagon and carry it up. You'd give out toting that rice up there. . . . It heavy. It's green, it's heavy. . . . Carry it to the end, get the wagon and load on it. You couldn't cut it all at one time. Be two or three weeks before you get your rice—a month before you get it all.
Q. The reason you'd stop cutting is because you didn't have enough place to hang it?
A. Enough place to hang it, and didn't had enough sheet to hang it over. . . .
Q. How did you make the frame that you've got the rice hanging on?
A. . . . Make a frame out of boards or either go in the woods and cut you a pole and make you a frame out of. [The frame was one top horizontal pole supported by upright poles, about six feet tall.]
Q. Every how often would they have a [supporting] pole?
A. Every ten foot. Some every five foot, depends on how you push the rice and stack it back. . . . Got to put something to hold it up. . . .

Then you get ready to hang it up . . . you have a crosspiece. Hang the rice up with the head hanging down. Spread your sheet down on the ground. . . . Then you got to put blocks up under there so when the wind blow it, it wouldn't blow the rice over on the ground. Just put them—a log and lay the sheet down, tack a nail here, a nail here, a nail here; and hold it. [He is saying that to prevent the rice from blowing off the sheet, a raised edge was made by placing small logs under the edge of the sheet.]

See it [the heads of the rice are hanging] about that high [three feet] over the sheet. . . .

If sun come out hot, in three days it start to dropping off.
Q. Within three days some of it is going to drop down on that sheet?
A. Drop down on that sheet.
Q. And you've got that sheet attached to logs on the ground?
A. On the ground so when the wind blows, it won't blow the rice on the ground. . . . Then you go there, you get that off. Sometimes you get two sacks full, sometimes three sacks full. Then you take that out. Take that out and go back tomorrow and do the same thing. . . .

. . . Then when you go there, it going to dry off, [if] you don't want to wait until it fall, take you little stick and hit you off some. Shell it like that if you want. . . .
Q. How you go about beating the rest of it off [that doesn't fall off by itself]?
A. . . . Let it stay up there about a week. Then you take you a stick

and hit the head of it. [Here the interview is confusing because he describes two different ways to get the last of the rice to fall. He has just described method one: hit the sheaves with a stick while they are still hanging. Method two is to take the sheaves down and hold them and beat them against a horizontal pole that is three feet above the ground. Of this method he said:]. . . . You take it off. Whip it on the pole, all of it will come off.

Q. . . . I want to see if I understand. It's been a week maybe it's been hanging; you figure it's dry; it's ready to get it all off. You then take a pole shaped like a "T" . . . got a crosspiece on the top . . . sitting how high above the ground?

A. Sitting about three foot above the ground. . . . I got a bundle and you got a bundle and you take it in your hand and whip it like that across there. . . . All of it come off. . . .

Hit that and that all of it come off at one time. Then you got mules or cows, then you throw the straw over to the cows and mule. And they'll eat that.

Q. What is the next stage now? You got the rice all off the stalks. What do you do next?

A. Move it and throw it over in the cow pen, or the mule pen; they'll eat that.

Q. The straw—but how do you keep preparing the grains of rice so you can eat them?

A. Put them in a sack—already dried—and put them up until you get ready to eat them. Give them sack—sack hold all you can. You can't eat it all.

Q. How do you clean the rice?

A. Clean it? With that maul like I show you.[2]

MATTHEW WILLIAMSON

Born: 1911
Description: large, reserved, distinguished-looking
Interviewed: May 14, 1985, on site of grandfather's and father's rice fields
Rice growers: Robert Gregg, Tony Howard, and Sherman Williamson
Method of rice cultivation: wetland
Period of rice cultivation: from ? to about 1922

[Matthew Williamson led me to the site of the rice field of his grandfather, Tony Howard. The field was beside a ditch on low land that formed the water course from a bay to Middle Branch Canal. Williamson recalled from his childhood that the area, now covered by large hardwood trees, was a rice field 60 feet by 300 feet.]

Q. Tell me about the rice field.
A. The rice field was . . . the north side of the ditch. And, of course, the ditch purpose was to furnish water when it got dry. They ran the rows parallel to the ditch so they could pour the water, and the furrows of the rice field would hold the water. . . . The rice was a way of life at that particular time, because they made it and of course they would use their old method of thrashing the rice after they would harvest it. . . . They had a reef hook. . . . And they would take some of the strands of the rice stalk and tie it in bundles. And after they stack it and let it cure out, they used what they call a mortar and a pestle. . . .
Q. Did they plant the rice on top of the row or in the bottom of the furrow?
A. They plant the rice in a furrow like—then they had another furrow so it would hold the moisture. . . .

[Speaking of watering the rice.] They had several places dug out along the ditch by the rice field. . . . You could go down there and take your bucket with a rope on it and throw it in the ditch and then pull it back to you and take the water and pour it in the field. . . .
Q. How deep would they try to get the water?
A. . . . I guess six inches—a foot.
Q. When you think was the last year they planted it?
A. I think my uncle, Robert Gregg, planted a few years after he lived here after my grandfather passed. . . . Must have been 1920 or 1922 was the last year they plant the rice.
Q. Do you have any idea when they started planting this rice field?
A. No, I don't. . . . I know my father [Sherman Williamson] planted a little rice up there on his farm near the ditch. He had one or two rows. I remember vividly working in it. We had a spring there in a ditch, and we would pull water from that ditch and pour it in the field . . . when it got dry.
Q. What years you think he planted?
A. . . . Let's see, that must have been about 1922. I believe it was. I believe I was round about seven years old.

Q. Did it stay pretty wet?
A. At that time, we had seasons when it was so wet that you didn't have to worry about pouring or dipping water from the spring. We call it our spring because we had a dug-out place there, and it was a spring at that particular time, because it would never go dry.

When it got dry, we had to dip water and pour it, but mostly this ground was very dark and sobby like. It's a good place for rice.
Q. How would you dip the water?
A. We had old wooden buckets, and some people had gourds during that particular time which they grew and they would use those as a dipper also.[3]

[Williamson showed me the site of his father's field and pointed to the spot where he had dipped water from the ditch when there was insufficient rain. The field was within about one hundred feet of the front door of Sherman Williamson's house.]

Hester Waiters

Lived: 1904–1988
Description: thin, cheerful, with musical voice, confined to wheelchair
Interviewed: August 14, 1987, at her home on a quiet street in Florence, S.C.
Rice grower: her stepfather, Horace Gregg
Method of rice cultivation: dryland
Period of rice cultivation: one year, about 1916

Q. Your mother and stepfather planted rice?
A. Yes. Plant it for one year.
Q. They didn't plant it but one year?
A. One year. After that they . . . get more rice coming in, you know, like you buy now. . . .
Q. What year were you born?
A. I born in 1904. My mother say.
Q. 1904? And how old you reckon you were the year they planted rice?
A. Well, I ought to be around about—Let me see, I know I ought to

be—Yeah, I know I ought to be about eleven or twelve years old. In them days.

Q. And you remember it good?

A. Oh yes, because I helped Papa plant it. Momma put me and my stepbrothers out to plant it.

Q. And where was it in relation to the house?

A. To the house, it was just along in there, oh, the place was so flat. And we planted it not far from the house. Not too far from the house.

Q. Was it a low place?

A. Yes, a low place. It grow better in bottoms.

Q. . . . Was it right with the rest of the garden or was it off by itself?

A. It off by itself, just like you plant a corn field. Crops off from the house.

Q. Was it in the woods or out in the—

A. No. It was out in the open. Land out in the open now.

Q. . . . Could you show me, if we—I can't go this trip, but sometime when I'm here, would you go down there and show me where you planted rice?

A. Where we planted at, it done change up so, and be full up. I know the spot of ground all right, but it done change up and the old grape arbors and pecan trees and all that gone. But I know that place when I go down there.

Q. How wet was it?

A. Well, it be real wet in them days. . . . It take a lot of water for them.

Q. And would you put any more water on it with buckets?

A. In the bucket? Well now, when you plant it you plant it in hills. About that far. And when it grow up, it grow up in big bunches. You work it around so it will stay like that. And then you gather it. You get that—Used to have an old-fashioned hook with a handle on it like that [a sickle]. Well, when we gather it, we go there and hold the heads, we bunch it together. Then you whack it, down low to the ground.

Q. Did they use that hook for anything else besides cutting rice?

A. You mean with that hook? Oh yeah, they'd cut broom straw—what would sweep you out the floor with. And that's as far as I know. . . .

Q. Did you plant it on top of the bed or down in the bottom of the bed?

A. We plant it kind of down in the furrow; then works the dirt kind of round it where it come up. Yeah. Just work your dirt up round it. Let the bed cover it after it come up.

Q. . . . And it would be real wet?

A. It grow better. It call for a lot of water.

Q. But you wouldn't pour any water on it?

A. No. Just the soil heavy. Now, light soil don't grow rice. Heavy soil. Yeah. Then when you gather it, you take cord string and tie it in bunches, in big bunches. And stack it in the field, up there. The heads be up, until you haul it to the house. And put it in your barn and put it in a dry place where it won't rain on it.

Q. How would you get it off the stalk?

A. Now, in them days, why, wasn't much material like there is now. We would take it and have a tin tub or a big dishpan, and with our hand then they'd strip it off in the pans. . . .

Q. And then what would you do with it next?

A. Well, in them days, my Uncle Bennie Bailey was living and he was growing rice, and on the weekends, me and my brothers, we'd go up there to thrash it off. He had a big block . . . like a saw-off a big block off a big tree. Then he had a big old pole, about as big as a baseball. [The hole in the block was as big as] an old-fashioned wash basin. . . .

Go down in that block. Then they had them something they call a mortar. . . . And I would take shucks. Corn be shuck off of cob, the dry shucks, and cut the end of them off, and spread on top of that rice . . . and then beat that. . . .

Q. And the shuck was on top of the rice?

A. Yes. To keep from breaking it up. Cracking it so bad.

Q. Would you have the shucks on the whole time you were beating it?

A. Yes. You put the shuck on to keep from breaking the rice too quick. And then the shuck wide. It bes wide all over it. It won't crack the rice. . . . It will help get that husk off it. . . . And then one of us get tired doing it like that, turn into the other one.

Q. Two of you be working at one time?

A. Just one.

Q. Then somebody else would take a turn?

A. Yeah. When you get tired, the other one; he get tired, then you go back. Then after a while, when you think you got all the chips up off of it, you take that chaff up off of that rice and throw it out. . . . Then take the rice, it be getting clean, then beat it again. Beat it until you get all that husk off it. . . .

You take it out in the pan. Blow all the [she makes a blowing sound] all the husks off it, we call it. . . . Put it back in there. You want to put you some more extra shuck on it until you get it clean. . . . Now the more you beat it the whiter it will be. If you didn't clean it as clean as you can get it, when you go home, your mother has to pick all that out.
Q. Would she pick on you too for not getting it clean?
A. Well yes. You'd get a whipping if you didn't clean it. . . .
Q. You all planted it one year? You remember more than most people remember that planted it for twenty years. I think you planted it more than one year.
A. Oh, yes. A lot of people planted it a heap of years, but I didn't know how many years at that time. I was telling you the year what—how we plant it. . . . One year. . . .
Q. When you planted—you're telling me a different way to plant the rice. A lot of people say they sowed it like oats. You didn't—?
A. No.
Q. You all dripped it in a hill?
A. [With disbelief.] Sowed it like oats? Then you'd gather, it all mixed up.
Q. Yeah. You'd have it in neat little—and it would be about a foot apart?
A. Yes. Be about a foot. Hoe it good. . . . And keep the strand out of it. Plow it. The more bed you put to it, the more it grow and get more strength to stand straight when it gets to storming. . . .
Q. And when you would beat it off, what would she [Mrs. Waiters' mother] keep it in?
A. . . . We'd just beat about enough for two or three messes at the time. . . . No more . . . until the next weekend you go fix some more.
Q. You'd do it on the weekend?
A. Yes.
Q. Everybody tells me that. How come you did it on the weekend?
A. Well, that's the best you did because you had to be working trying to grow other things.
Q. Yeah. On Saturday?
A. You know, all through the week until Saturday. And Saturday would be our main time to prepare something for Sunday.
Q. How many days—times would you eat rice?
A. We eat rice? We would eat rice every day we could get it if you let—[laughing].

Q. And your momma would keep it before you all had husked it—what would she keep it in—a crocus sack or what?
A. Well, we wouldn't take it off the stalk until about that Friday. . . .
Q. Tell me about your Uncle Bennie Bailey that raised rice. He raised rice a long time?
A. He raised for a long time. . . .
Q. Now where did he grow his rice?
A. Well in them days I didn't know where the Blanche and them grow their—where they grow it at. But I know they had the concern for to beat the rice off at his house. . . . Ask him could I come. "Uncle Bennie, can I come up—come and beat us rice off?" He say, "Yeah, come on. . . ."
Q. . . . When he wanted to get his rice off the stalk, he pulled it off with his hands, the same way you did?
A. I believe he did. Because that how I learned.
Q. And you'd go and use his mortar and pestle, but you had one at your house?
A. No, I didn't had none. We used to go use his.
Q. Because you just planted it one year, so when you wanted to beat it off, you'd have to take it to his house?
A. And that the onliest house I knowed.
Q. That had a mortar and pestle?
A. Yeah.
Q. How far from your house was his house?
A. I know Uncle Bennie was good a mile. In them days.
Q. What did you take it over there in, a . . . ?
A. They'd put it in a sack . . . what don't have a hole in it . . . and put it on our back. . . .
Q. And you don't have any idea where he planted his?
A. No, I couldn't even know where he planted his because I wasn't running around on his property.
Q. But he owned his own land?
A. Yes. Work hard.
Q. Did you know anybody else that grew rice?
A. In them days, children couldn't get around to go from house to house.
Q. Uhhu. You stayed right at home.
A. Yes.

Q. Worked when they told you to.
A. Yes. Going right when I come in from school, plenty work.[4]

FRANCES MISSOURI JOHNSON

Lived: 1915–1991
Description: thin, agile, bespectacled, quick-witted
Interviewed: March 12, 1984, and July 2, 1984, at her home in Mars Bluff
Rice grower: her father, Vico Johnson
Method of rice cultivation: dryland
Period of rice cultivation: from ? to about 1939

Q. Do you remember anybody planting rice down here?
A. My father used to plant rice. . . . Used to plant rice and beat it in a mortar. Had a mortar and a pestle.
Q. Where did he plant rice?
A. . . . Right down here, a little piece over there in the field.
Q. Was it in the dry field or down in the swamp?
A. No, it's the bottomlike. You know, black soil. . . .
Q. Did it stay wet or did it stay dry?
A. . . . It's according to how it rained. And then he cut it. When it get dry, he'd take a reef hook and cut it . . . and then tie it in bundles. Then he'd bring it to the house and make something like a flat and put it on there and get something like a frail bars [flail] and I think, I believe, you burn them, then you twist them. And you put the rice on this scaffold and the sheet under there. And then you beat it, and then you take it to Lake City and have it thrashed [husked].
Q. Did you ever help him in the rice?
A. Sure did. Us used to help him beat it off. . . .
Q. . . . I want to start at the beginning. . . . How would you put the seed out?
A. Drop it.
Q. Would you put them in hills?
A. Yeah, in hills.
Q. How far apart?

A. About like that.
Q. About a foot apart. And it would be on top of the bed or down—
A. . . . You drop it in a furrow, and then he'd come along and cover that rice up. . . .[5]
Q. Would it be pretty wet when you'd be out there putting the seed in?
A. Sometimes it would. . . . Sometime it would get pretty wet, and sometime it wouldn't. It's according to how much it rained, because he had it down there in a stiff place. . . .
Q. And he would cut it with a—?
A. Reef hook.
Q. And how did he bring it up to the house?
A. . . . I think he haul it with the mule and wagon. I don't think us tote it.
Q. Where would he put it to dry?
A. . . . Put it on something like a rack. . . . and let it dry. . . .
Q. Then it would be a different scaffolding you'd put it on to beat it?
A. Unhu. . . . you make it kind of like a scaffold. . . . Something like a flat, you nail it on there. And then you put your rice on here and put the sheet down here on the ground. But you got have . . . wide enough apart when you beat it for the rice to go through the cracks, down here on the sheet. . . .
Q. What's that thing made out of that's got cracks in it?
A. It just little [half-inch] poles. I reckon they are oak or hickory or something. Little poles, and got them nailed across, just like that.
Q. How far apart are they?
A. . . . About that far apart [about eight inches].

You put it [the bundles of rice] on the scaffold. . . . It's right down flat . . . [just over three feet above the ground]. And you beat it. And you let it [the rice] go through the cracks down here on the ground on the sheet. And then when you done beat it, you get that—you move all that trash out the way. And you go under there and pull your sheet out, and get all that trash out the rice, and you put it in a bag and take it to Lake City [to a mill to be husked]. . . .
Q. How many people would beat it at once? . . .
A. . . . I believe about two could beat at a time. . . . One on this side and one on that side. And you burn that thing and you twist it. Twist it like that, so it would flop kind of like a chicken head when you wring it. . . . It would flap.

Q. What do you mean—you'd burn what?
A. Burn this thing so you could twist it. Burn it. When you cut that pole to beat the rice. It'll be little something about that big, and you burn it and get it where you could make it soft so you could twist it. And beat the rice.
Q. Nobody ever told me about that. . . . Tell me . . . What kind of pole would they take? . . .
A. . . . I think it would be hickory. . . . Little hickory. It be about that big [one inch or a little larger]. . . .
Q. And they'd take the hickory tree [sapling] and they'd burn it? What part of it?
A. Burn it a piece about that long from the butt down here. And then when they get it burned where you could twist it, take it and wring it. Wring it just like that, and you'd hold the small end and beat the rice. And the big end would be down there. See, every time you bring it up, it would twist round like that and you go back down.
Q. O.K. You take a piece of hickory, about how long would it be?
A. I don't know. It wasn't quite long as this chair [three-seat sofa].
Q. So that would be about four or five feet long. And then about a foot and a half back from the end of it you'd burn it?
A. Yeah.
Q. . . . When would you twist it?
A. After it cooled.
Q. And how much would the burning take off of it?
A. . . . You just burn enough so you could twist it. . . .
Q. And then when you twisted it, it would stay twisted?
A. It would stay twist, because you be beating the rice with it. Every time you bring it up. You see you come down on the rice like that. And you bring it up, that twist will go round just like that, and you carry it back down.
Q. And when you beat—What did you call that . . . hickory [flail] that you swing around and it comes down?
A. . . . A frail bar. . . .
Q. Frail bar? B A R?
A. I reckon so. . . .
Q. . . . Do you remember using the mortar and pestle to husk the rice?
A. I do. I used to beat—oh, I used to beat rice.
Q. How much would you put in there at once?

A. I reckon about three or four quarts. . . .

See, you take that pestle and come down in there just like that. Little in the middle—it got to be like that; you can put both hands around it. And come down with the pestle. Come down in there. . . .

When one get tired, another one take it. That's the way you had to do it. One just couldn't stay there and beat. Our brother used to help us beat too. . . .

Q. You'd stop and get rid of the husks—

A. . . . You take it up and you fan it.

Q. How you fan it?

A. You've got to put it in the dish-pan and take that dish-pan and do it just like that [makes a motion that would cause the rice to jump up and down in the pan], and all that trash will jump out on the ground. . . .

Q. So, when you do the dishpan like that, that makes the rice kind of jump up in the air. And you holding the dishpan flat?

A. No, you hold it to the side like this . . . and that stuff will go on the ground. . . .

Put it in a dishpan and fan all that trash out it. Then you got to put it back and you get it good and clear, then you get some shucks. . . . Corn shuck. . . . and tear them up, in little strips and throw it in the rice, in the mortar. And then take that pestle and go back at it again. . . . to clear the rice. . . . Make it pretty. . . . Make it white. I used to do that more times than a little.

Q. So you would have finished beating the husk off, and then you'd put those corn shucks in—strips of corn shuck?

A. Uhhu.

Q. And beat it some more?

A. Until you get it right. . . . Put it back and do some more. . . . If you don't have it right, have to put it back in there, because Momma would make you get it right. . . .

. . . Take the last us plant was to Lake City. . . . He'd have to get somebody to take him in a car.

Q. About how many bags would he have?

A. Well, it according to—I think he would have about two or three or something like that.

Q. . . . About what year you think that was that he started going to Lake City with it?

A. I don't know. . . .

Q. All while you were a child, you all were beating it with the mortar and pestle?
A. Until us move over here [1937 or 1938]. Because us used to plant it over yonder [on the nearby farm where they lived before her father bought his own farm], beat it in the mortar. And after we move over here, beat it in the mortar. Then he start to Lake City.
Q. Over there, what kind of place did he plant it on?
A. In the low place. . . .
My father . . . die in '47.
Q. What year do you think he quit planting?
A. I believe it was in about eight or nine [years], I guess before he passed. . . . It could have been a little longer; it could have been a little less. But I know he stopped planting before he passed.
Q. . . . And why did he quit?
A. . . . I don't know. . . .
Q. Did he get in too poor a health to be farming?
A. Well, us farmed right on because he had two boys. I don't know why he quit either. But I know he quit planting rice. . . .
Q. Spell your father's first name for me.
A. V I C O.
Q. You know where he came from?
A. Right down here. It was his home here.
Q. And his mother and father, you know where they came from?
A. They was born right here, I reckon. I reckon, I don't know. I think they was. His mother was Mindas, and his daddy was Daniel.
Q. Spell his mother's first name?
A. I don't know how you spell Minda. . . .
Q. . . . Would your father ever have neighbors who would help him growing the rice?
A. No. You see he had eight head of children and his children would help him.
Q. Would he ever sell rice to anybody?
A. No. . . .
Q. Would he ever give it away?
A. No. . . .
Q. Did he raise just enough for the family?
A. Yes. . . .
Q. Would it get you through the year every year?
A. Yes.[6]

Ida Ellison Zanders

Lived: *ca.* 1912–1992
Description: slowed down physically, but mind and spirit still strong, spoke in joyous tone
Interviewed: August 4 and 7, 1987, at her home in Florence, S.C.
Rice grower: her mother, Fanny Jolly Ellison
Method of rice cultivation: dryland
Period of rice cultivation: from ? to 1930s

Q. Do you remember any of those old-timers planting rice?
A. Unhu. My momma plant rice right there on Palmetto Street [a main thoroughfare in Florence].
Q. You're kidding.
A. Fanny plant rice. Made me and my husband—the same man, I was courting him then—used to have to go out there and get it and beat it.
Q. In the big mortar and pestle?
A. Unhu.
Q. Did she bring that mortar and pestle from Mars Bluff into town?
A. Unhu. Unhu. . . .
Q. How big around was it?
A. It wasn't no big one. It was about like that.
Q. That's about two feet or better across. And she would make you go out there and beat it while you were courting?
A. Whilst I was courting. If he come by there in the evening, "Ollie, you and Buck go out there and get that rice." Well, I had to do it then. . . .

[Tape off while a woman comes in with some peaches for Mrs. Zanders.]

Q. You were telling me about raising rice on Palmetto Street.
A. That's 1109 Palmetto. There's a field from that house to the next house. . . . They used to tell my momma to plant it to keep it from growing. And she plant rice, planted okra. Plant anything in there.
Q. Would she have her vegetable garden there?
A. Right there.
Q. Would she plant the rice right next to the vegetables?
A. Unhu. I believe she plant the rice first. But I know I's beaten major rice right there. Time she say—I stop him from coming to see me [laughing]. . . .

Q. She'd stop him from coming?
A. I stop him. . . . I got tired beating rice.
Q. Oh. . . . She'd just make you do it if he was there?
A. If he was there, because he was a boy.
Q. Would you all both beat it together or one at a time?
A. He'd beat it most for me [hearty laugh].
Q. You wouldn't have two of those pestles?
A. Unhu.
Q. You'd both be beating at the same time?
A. I'd just would be making time.
Q. You'd just be making time, and he would be doing the husking?
A. Unhu.
Q. Did your mother have rice out at Mars Bluff?
A. . . . Oh, yeah. Because my momma didn't buy nothing.
Q. Where do you think she raised the rice out there?
A. Somewhere behind there. Had a big sycamore tree in the yard. . . . Behind the sycamore tree. . . .

[A few days later in the cool of the morning, Mrs. Zanders and I drove out to Mars Bluff so she could show me where her mother's rice patch had been. At the center of Francis Marion University campus, we stood on the spot where her mother's house had been.]

A. Well, go round the house, go out the back door. Come round the house and go that way. The rice was over here. And the beans and the peas, the collards and the peppers.
Q. . . . Was it a lower area than—?
A. Yeah. It was black dirt.
Q. Was the whole garden black dirt or was—?
A. The whole garden.
Q. And the rice was on a lower part or not?
A. On the lower part. Oh, you better not walk me too fast; I give out. . . .
Q. O.K. We'll stop and rest a minute.
A. Oh, ha. The Lord got patience. Got patience with me, because ain't through with me yet—and when you get to that tree you go about right through there. That the gardens, the garden spot . . . Get all the okras you want. . . . Everything was planted and then the rice come over there last.

Q. Would there ever be any water standing in the rice?
A. No.
Q. It was dryland rice?
A. Unhu. Us ain't never in no water. Because you know she wouldn't let us play—scared of snakes.

[On the way back to Florence, Mrs. Zanders told how they flailed the rice.]

Q. Can you remember up in Florence, when your mother would harvest the rice, how would she do it?
A. That's where they'd cut it down and put it in that thing and beat it. Put it in a sack.
Q. How would you get it off the stalk?
A. Us would cut the stalk down and put the stalk in a sack. And beat it in a sack. And when you beat it so much and the rice come out, then you take the rice out.
Q. Well, that's the smartest idea I ever heard. . . . In a crocus bag?
A. Yeah. Beat it until the rice come out it. Then you take the stalks out. . . .
Q. When you got the rice in the sack and you're trying to beat it off the stalk, you pick up the whole sack and you hit the sack against what?
A. Take that maul and hit the sack.
Q. You just beat that maul against the sack and then you take the straw out, and the rice stays in the bag.
A. Uhhu.
Q. Did you ever see anyone else do it that way?
A. No. Us didn't go. You know, they didn't carry us around other people. They'd go, go to your house and set down and talk. . . .
A. [Later she tells about how they winnowed the rice.] . . . And then you beat it and sift it, they call it, up and down. For the wind to blow it out.
Q. What would you drop it into when you were sifting it?
A. In another sack.[7]

Notes

Introduction

1. Lawrence W. Levine, *Black Culture and Black Consciousness: Afro-American Folk Thought from Slavery to Freedom* (New York, 1977), xiii.

Chapter 1

1. Archie Waiters, interview, June 26, 1984, transcript in author's notebooks of interviews, 202. All interviews were conducted and recorded by the author unless otherwise noted. The author's tapes and transcripts will be deposited in the Caroliniana Library, University of South Carolina, Columbia, S.C.

2. The year of Gregg's birth was originally calculated from information given by Catherine Waiters. She said that he was ninety-three when he died in 1938 (Catherine Waiters, letter to author, November 11, 1985. All letters to the author are in the author's possession unless otherwise noted.). In 1991, Gregg's great-grandson learned from the county health department that Alex Gregg was born in 1844 and died in 1938 (Otis Waiters, letter to author, October 29, 1991).

3. A. Waiters interview, June 5, 1985, pp. 727–28, 734, 736, 738.

4. Peter Wood, *Black Majority: Negroes in Colonial South Carolina from 1670 Through the Stono Rebellion* (1974; rpr. New York, 1975), 333–34, 341. Wood wrote of the relatively small number of Africans who entered at Georgetown from 1735 to 1740: "The Treasury Records omit . . . those Africans perhaps several hundred who arrived at Georgetown and Port Royal" (334). Treasury records show that 11,562 Africans entered Charleston during the same period. George C. Rogers, Jr., wrote: "Slaves were not imported into Georgetown directly. Duties were paid on slaves brought into Georgetown in the years 1755, 1764, 1765, 1769, 1771, 1772, 1774, but these were in small lots. . . . The largest

number of slaves imported was 112 in 1774. . . . The overwhelming number of slaves brought into Georgetown were bought in Charleston" (George C. Rogers, Jr., *The History of Georgetown County, South Carolina* [Columbia, S.C., 1970], 53). Charles Joyner concurred that most inhabitants of the Georgetown area transacted their business in Charleston (Charles Joyner, telephone conversation with author, March 13, 1990).

5. Alexander Gregg, *History of the Old Cheraws* (New York, 1867), 68. Robert Mills wrote that most of the settlement of the Marion area occurred about 1750, "chiefly by Virginians" (Robert Mills, *Statistics of South Carolina Including a View of Its Natural, Civil, and Military History, General and Particular* [Charleston, S.C., 1826], 622). Robert L. Meriwether wrote that the Williamsburg area of South Carolina was surveyed in 1732. Because the government opened the area to claims by people who were not settlers, the settlers had trouble finding land (Robert L. Meriwether, *The Expansion of South Carolina, 1729–1765* [Kingsport, Tenn., 1940], 79–81, 90). G. Wayne King wrote: "Established in 1770, Hopewell Presbyterian Church at Claussen [about a half mile from Mars Bluff] had its origins in the Williamsburg settlement" (G. Wayne King, *Rise Up So Early: A History of Florence County, South Carolina* [Spartanburg, S.C., 1981], 14–15).

6. Rogers, *The History of Georgetown County,* 343. Rogers wrote: "In the fifty years from 1810 to 1860 the slaves composed 85 percent to 89 percent of the population of Georgetown District" (343).

7. Lorenzo D. Turner, *Africanisms in the Gullah Dialect* (1949; rpr. New York, 1969), 1, 3. Turner defined the region where Gullah was spoken as extending "along the Atlantic coast approximately from Georgetown, South Carolina, to the northern boundary of Florida. It is heard both on the mainland and on the Sea Islands near by" (1).

8. Daniel C. Littlefield, *Rice and Slaves: Ethnicity and the Slave Trade in Colonial South Carolina* (Baton Rouge, 1981), 113. Joseph E. Holloway emphasized the importance of the 32 to 40 percent of Bantu provenance; because of the homogeneity of Bantu culture, Bantu was the most important cultural influence in South Carolina (Joseph E. Holloway, "The Origins of African-American Culture," in *Africanisms in American Culture,* ed. Joseph E. Holloway [Bloomington, Ind., 1990], 4–9). Joe Opala suggested especially strong linguistic connections between South Carolina and Sierra Leone (Joseph A. Opala, *The Gullah: Rice, Slavery, and the Sierra Leone–American Connection* [Freetown, Sierra Leone, 1987], 15–20).

9. Jacob Stroyer, *My Life in the South* (3rd ed.; Salem, 1885), 9.

10. A. Waiters interview, March 23, 1984, p. 102.

11. Louise Daniel Hutchinson, *Out of Africa: From West African Kingdom to Colonization* (Washington, D.C., 1979), 22–23. Mariah Malinka's last name might have been spelled *Malinke.* I was the one who gave it the spelling *Malinka.*

12. Nehemia Levtzion, *Ancient Ghana and Mali* (London, 1973), 73, 208–209; J. D. Fage, *A History of West Africa: An Introductory Survey* (4th ed.; Cambridge, Eng., 1969), 21–24; George Peter Murdock, *Africa: Its Peoples and Their Culture History* (New York, 1959), 73.

13. Levtzion, *Ancient Ghana and Mali,* 73–74, 95–97; Michael Mullin, ed., *American Negro Slavery: A Documentary History* (Columbia, S.C., 1976), 48.

14. A. Waiters interview, July 31, 1987, p. 1147.

15. Melville J. Herskovits and Frances S. Herskovits, *An Outline of Dahomean Religious Belief* (Menasha, Wis., 1933), 56–58; Melville J. Herskovits, *Dahomey: An Ancient West African Kingdom* (1938; rpr. Evanston, Ill., 1967), II, 248–55. Mechal Sobel defined *Da* as the name used by the Fon of Dahomey for "a force which manifests itself in the world in a number of ways" (Mechal Sobel, *Trabelin On: The Slave Journey to American Afro-Baptist Faith* [Westport, Conn., 1979], 13). Charles Joyner, writing about the perpetuation of African religion in the Waccamaw River area, said: "High in the pantheon of African deities, the snake god of so many African cultures—the Ewe, Fon, Bantu, Dahomey, Whydah, Yoruba—symbolized the cosmic energy of nature, the arbiter of fortune and misfortune" (Charles Joyner, *Down by the Riverside: A South Carolina Slave Community* [Urbana, Ill., 1984], 145).

16. Herskovits wrote that Da was responsible for the mysterious way that one's fortune changed; therefore Da must be watched and placated. "This vigilance is, however, the responsibility only of those in power—heads of households, but not women and minors" (Herskovits, *Dahomey,* II, 251). It may be significant that Waiters and Gregg were both heads of household. Prior to December, 1992, I had never heard anyone except Waiters use the expression; then Jane Vernon phoned from Mars Bluff to tell me that she had heard Mary Scott Cooper say "Great Da." When I spoke with Mrs. Cooper, she explained: "That's just a word that we Black people use. . . . Somebody says, 'I wrecked my car last night,' and you say, 'What?' He says, 'I ran off the road,' and you say, 'Great Da'" (Mary Scott Cooper, telephone conversation with author, January 24, 1993).

17. Discussing the use of the names of gods as a way of tracing African origins, Herskovits wrote: "The coastal area of the Gold Coast was occupied by the Fanti tribes, who because of their control of this strategic region, acted as middlemen for tribes to the north. Yet all evidence from recognizable survivals such as the many Ashanti-Akan-Fanti place names, names of deities, and day names in the New World are evidence that the sources of the slaves exported by the Fanti were in greatest proportion within the present boundaries of the Gold Coast colony" (Melville J. Herskovits, *The Myth of the Negro Past* [1941; rpr. Boston, 1972], 35). Herskovits' implication here is that survivals of deity names might be used as evidence of the place of origin of the Africans who used them.

18. Alex Haley, *Roots: The Saga of an American Family* (New York, 1976).

19. Charlie Grant, interview by H. Grady Davis and Lucile Young, pre-

pared by Annie Ruth Davis, August 31, 1937, in George P. Rawick, ed., *The American Slave: A Composite Autobiography* (1941; rpr. Westport, Conn., 1972), Vol. II, Pt. 2, pp. 172–73. Grant was listed as eighty-five years old in 1937, so he was probably about seven years younger than Alex Gregg. He lived on Dr. William Johnson's farm at the north edge of Mars Bluff.

20. Philip M. Hamer and George C. Rogers, Jr., eds., *The Papers of Henry Laurens* (Columbia, S.C., 1968—), IV, 281–82. A letter written in 1755 from Henry Laurens to London, England, spoke of selling slaves to people in Williamsburg: "We shall have a great deal [of indigo] offer'd to us from such Persons as deal with us for Slaves from Williamsburg Township" (*ibid.*, I, 279).

21. I have chosen to use the word *slaves* even though I object to it as the primary identification of any people. The more accurate phrase recommended by Paula S. Rothenberg, "African American people who were enslaved," is too long and distracting for repeated use (Paula S. Rothenberg, *Racism and Sexism: An Integrated Study* [New York, 1988], 273).

22. A. Waiters interviews, February 20, 1982, pp. 98–99, June 5, 1985, pp. 712–13, June 6, 1985, pp. 812–13. (Here and in some other quotations of interviews, I have put together quotations from separate interviews to give a more complete account.) Gregg also told Waiters of a large building on the farm at Mars Bluff where the slaves lived together before they built their hewn-timber houses, so their being panicked by the light at the window could have occurred at Mars Bluff. I believe, however, that it must have happened when a large number of newly arrived Africans were housed together in Charleston. At Mars Bluff, slave owners usually bought a few slaves at a time. It is unlikely that there would have been forty or fifty new slaves together before any houses were built or that the fence around the building would have been ten feet tall.

23. Evan Pugh, "The Private Journal of Evan Pugh," as condensed and typed by Mr. and Mrs. O. L. Warr, 58 (Typescript in possession of Darlington County Historical Commission, Darlington, S.C.). I am indebted to Horace Fraser Rudisill, who is editing the journal for publication, for informing me of its existence.

24. [Charles Ball], *Slavery in the United States: A Narrative of the Life and Adventures of Charles Ball, a Black Man* (1836; rpr. New York, 1969), 71. No author is given for this book in the book itself. Library records show the author as Charles Ball. Because the book is written in first person, the implication is that it is an autobiography. But page one refers to a Mr. Fisher as the author, and another statement explains that the story is as told by Ball but the wording is not his.

25. *Ibid.*, 128–29.

26. "The distance from Georgetown by water . . . to Mars Bluff [is] 100 miles, by land 65" (Harvey Toliver Cook, *Rambles in the Pee Dee Basin, South Carolina* [Columbia, S.C., 1926], I, 57).

27. Joyner, telephone conversation with author, March 13, 1990. The papers of R. F. W. Allston reveal his reluctance to sell any of his slaves, though he sold slaves for others when handling estates. During the Civil War, Allston was obliged to move his enslaved workers to North Carolina, just twelve miles above Cheraw, South Carolina, and there is a record of a sale of one family at that time (J. H. Easterby, ed., *The South Carolina Rice Plantation as Revealed in the Papers of Robert F. W. Allston* [Chicago, 1945], 29, 68, 189, 276, 384, 389). Rogers, writing about Allston selling slaves for estates, said: "Most of this buying and selling, however, was within the district" (Rogers, *The History of Georgetown County,* 329). Norrece T. Jones, Jr., wrote that after the 1780s few slaves were brought into the district of Georgetown and "apparently even fewer were exported to foreign parts than in other areas" (Norrece T. Jones, Jr., *Born a Child of Freedom, Yet a Slave: Mechanisms of Control and Strategies of Resistance in Antebellum South Carolina* [Hanover, N.H., 1990], 44).

28. Chapman J. Milling, "Mechanicsville," in *Darlingtoniana: A History of People, Places, and Events in Darlington County, South Carolina,* ed. Eliza Cowan Ervin and Horace Fraser Rudisill (Columbia, S.C., 1964), 148. Milling was born in 1901, so his observations date from the twentieth century. One record, the Bill and Receipt for Negro Nat, shows that in 1834 the Winyaw and Indigo Company, located in Georgetown, sold a slave named Nat to a man in Sumter, west of Mars Bluff (John Blount Miller Papers, Box 2, Folder 2, Manuscript Department, William R. Perkins Library, Duke University, Durham, N.C.). Such transactions between Georgetown and the pine belt were probably infrequent.

29. When transcribing the Mars Bluff interviews from tape to paper, I changed all words to standard English, so the interviews do not reveal the similarities between the Mars Bluff dialect and African languages or Gullah. For example, I wrote that Archie Waiters said, "They give that to you." He had said, "They ge that to you" (A. Waiters interview, June 5, 1985, p. 730).

30. Waiters added, "Old people, well educated—die off right away" (A. Waiters interview, March 9, 1990, n.p. Interviews taped after June 29, 1988, are not in written form, so they have no page numbers.) I interpreted this statement to mean that when Mars Bluff slaves went to Georgetown to search for the older, better-informed people from their ethnic group (the people who would be best able to tell them about new arrivals from their ethnic group), they found that those people had died soon after their arrival in South Carolina.

31. Herskovits, *Dahomey,* II, 64–65.

32. Meriwether, *The Expansion of South Carolina,* 94; David Duncan Wallace, *South Carolina: A Short History, 1520–1948* (Columbia, S.C., 1951), 365; Patrick S. Brady, "The Slave Trade and Sectionalism in South Carolina, 1787–1808," *Journal of Southern History,* XXXVIII [November, 1972], 612. From 1804 through 1807, 39,075 Africans were brought into Charleston (Wallace, *South Carolina: A Short History,* 365). John Hope Franklin wrote that there

was illegal slave trade after 1808 (John Hope Franklin, *From Slavery to Freedom: A History of American Negroes* [New York, 1947], 140, 182–84). I quote William Willis Boddie with some misgiving. He wrote: "In 1808, there was a shipload of Guinea negroes sold in Williamsburg, South Carolina" (William Willis Boddie, *History of Williamsburg* [Columbia, S.C., 1923], 332).

33. Julia Floyd Smith, *Slavery and Rice Culture in Low Country Georgia, 1750–1860* (Knoxville, 1985), 103; Stroyer, *My Life in the South,* 30. Smith wrote: "Between 1830 and 1860 Virginia exported approximately 300,000 Negro slaves" (103). Daniel M. Johnson and Rex R. Campbell wrote that while South Carolina was a recipient of slaves from the Middle Atlantic states, South Carolina was also a major supplier of slaves to the states to the south and west (Daniel M. Johnson and Rex R. Campbell, *Black Migration in America: A Social Demographic History* [Durham, N.C., 1981], 22–23).

34. [Ball,] *Slavery in the United States,* 127–28, 132, 36–37. Ball was born about 1781 and was a young man with small children at the time he was taken to South Carolina; thus I calculate that he was taken south very early in the nineteenth century (16, 36).

35. *Ibid.,* 68–69.

36. Clement Eaton, *The Growth of Southern Civilization, 1790–1860* (New York, 1961), 31, 37–38.

37. Sallie Gregg Wallace, conversations with author, Mars Bluff, S.C., 1930s and 1940s.

38. I am confident that the Flemings were purchased in Louisiana because my grandmother would not have described them as being from Louisiana if they had lived first in South Carolina and had gone to Louisiana with their owner. The orphaned girl was Sarah Edwards, who was born in 1812 and married J. Eli Gregg in Society Hill in 1830, according to the Gregg family bible. From these dates, I calculate that the Flemings moved from Louisiana to Society Hill about 1822. Louise Miller McCarthy wrote that Sara Edwards Gregg moved to Mars Bluff about 1836 (Louise Miller McCarthy, *Footprints: The Story of the Greggs of South Carolina* [Winter Park, Fla., 1951], 49). I assume that the Flemings moved to Mars Bluff at that time.

39. Carter G. Woodson, *The Negro in Our History* (5th ed.; Washington, D.C., 1928), 218–19.

Chapter 2

1. A. Waiters interview, July 31, 1987, p. 1122. Various African ethnic groups had different customs for naming people. The name *Rooster* might have been Gregg's personal name, or it could have come from a tradition like the Mandingo one, in which each ancient family had a clan name, frequently the

name of an animal (Turner, *Africanisms,* 31, 35, 36, 38; Sterling Stuckey, *Slave Culture: Nationalist Theory and the Foundations of Black America* [New York, 1987], 196).

2. I am deeply indebted to all of the former slaves who gave so generously of their memories; to Annie Ruth Davis, who was responsible for the Marion County, South Carolina, interviews; to George P. Rawick for making these valuable interviews so readily available in *The American Slave: A Composite Autobiography;* and to Greenwood Publishing Company, an imprint of Greenwood Publishing Group, Inc., for its generous sharing of this material.

3. A. Waiters interviews, July 25, 1984, pp. 321–22, 325, June 5, 1985, pp. 701, 712–13.

4. Seventh United States Census, 1850, Schedule 2, Slave Inhabitants in the County of Marion, State of South Carolina, 77[?], 91B, 295, 295B, in South Carolina Department of Archives and History, Columbia, S.C. I did a rough tally, of the slaves counted in this 1850 record, on several pages that had names of owners that I recognized as probably being in Mars Bluff. It was not a complete or accurate survey; I merely scanned a few pages to get some idea of the number of slaves living on a single farm. The numbers ran like this: three landowners owned one slave apiece, four owned three to five slaves, four owned ten to twenty slaves, three owned twenty to thirty, one owned thirty-six, one owned forty-nine, and one owned fifty-five. Of course, there were some farms at Mars Bluff that had many more slaves: Malachi Murphy, one of the earliest settlers in the Mars Bluff area, "is said to have given 100 slaves to each of three sons" (Gregg, *History of the Old Cheraws,* 72).

5. Ulrich Bonnell Phillips, *American Negro Slavery: Survey of the Supply, Employment, and Control of Negro Labor as Determined by the Plantation Regime* (1918; rpr. Gloucester, Mass., 1959), 156; Wallace, *South Carolina: A Short History,* 340–41, 363–64, 710–11. On 710–11, Wallace gives the following figures for the increase in the African-American population relative to the white population:

RATIO OF AFRICAN AMERICANS TO EUROPEAN AMERICANS

	Marion County	*Darlington County*
1820	3,549 to 6,652	4,552 to 6,407
1860	10,183 to 11,007	11,929 to 8,421
1880	18,226 to 15,881	21,556 to 12,929

(Figures are given for two counties because the Great Pee Dee River cut Mars Bluff off from most of Marion County, of which it was a part, and fostered its association with Darlington County to the west. Consequently, in economic and cultural development Mars Bluff may have been more like its neighbor to the west than to Marion County.) These increases came well after the invention of the cotton gin. Writing about the Pee Dee area at mid-eighteenth century, before the invention of the cotton gin, King stated: "Most whites did not own more than three or four slaves" (King, *Rise Up So Early,* 17).

6. A. Waiters interview, June 5, 1985, pp. 710–11, 718, 726.

7. Wood, *Black Majority,* 107.

8. A. Waiters interview, June 5, 1985, pp. 705–707, 716, 718–19. Grits were produced from corn specifically ground for cooking as a cereal. Mush was made of cornmeal, the same meal that was used for making cornbread. Both grits and mush were cooked in the same way that hot cereals are cooked today, but for a much longer time.

9. A. Waiters interview, June 5, 1985, pp. 707–708. Waiters' sister Maggie Waiters Egleton had also lived with her grandfather Alex Gregg. She gave a similar account of how her grandfather had described what they ate when they were slaves: "Used to pour trough in there and pour milk and mush—clabber milk. And they used to eat with stick spoons" (Maggie Waiters Egleton, interview, March 3, 1986, p. 1043). Leland Ferguson wrote about Europeans and European Americans using trenchers, small wooden trays from which two or more people ate. Although Ferguson's book is about early African Americans, the reference to trenchers spoke only of their use by Europeans and European Americans, a use that he says was abandoned in the eighteenth century (Leland Ferguson, *Uncommon Ground: Archaeology and Early African America, 1650–1800* [Washington, D.C., 1992], 97–98).

10. [Ball,] *Slavery in the United States,* 85.

11. *Ibid.,* 194–95.

12. The name *Pigiott* has six or more spellings. I have used the spelling found on the United States Geological Survey Map of 1938–40 marking the cemetery where Emma Pigiott Gregg is buried.

13. The term *the place* was generally used at Mars Bluff in lieu of *the farm.* Throughout this book, the term *place* refers to the land and buildings of one landowner.

14. A. Waiters interview, June 5, 1985, pp. 744–45, 735, 746.

15. "The value of a good farm negro in Williamsburg in 1800 was $500.00; in 1820, $725.00; in 1840, $800.00; in 1850, $700.00; in 1860, $1200.00" (Boddie, *History of Williamsburg,* 338).

16. Sylvia Cannon, interview by H. Grady Davis and Lucile Young, prepared by A. Davis, August 4, 1937, in Rawick, ed., *The American Slave,* Vol. II, Pt. 1, pp. 188–90, 192. The bracketed comment is the interviewer's. Mrs. Cannon was eighty-five years old in 1937, so she was born about 1852. She would have been thirteen years old at emancipation and younger than that when sold.

17. Grant interview, in Rawick, ed., *The American Slave,* Vol. II, Pt.2, p. 173.

18. Lizzie Davis, interview by A. Davis, December 21, 1937, in Rawick, ed., *The American Slave,* Vol. II, Pt. 1, p. 293. There is no record from Mars

Bluff of a slave's reaction to the traumatic event of an estate sale. Jacob Stroyer described one in Sumter County.

> A Mr. Manning bought a portion of them and Charles Login the rest; these two men were known as the greatest slave traders in the South, my sisters were among the number that Mr. Manning bought.
>
> He was to take them into the state of Louisiana for sale, but some of the men did not want to go with him, and he put those in prison until he was ready to start. . . .
>
> We arrived at the depot and had to wait for the cars to bring the others from the Sumterville Jail, but they soon came in sight, and when the noise of the cars died away we heard wailing and shrieks from those in the cars. . . .
>
> While the cars were at the depot, a large crowd of white people gathered, and were laughing and talking about the prospect of negro traffic; but when the cars began to start and the conductor cried out, "all who are going on this train must get on board without delay," the colored people cried out with one voice as though the heavens and earth were coming together, and it was so pitiful, that those hard hearted white men who had been accustomed to driving slaves all their lives, shed tears like children. As the cars moved away we heard the weeping and wailing.
>
> (Stroyer, *My Life in the South,* 42–43)

19. A. Waiters interview, June 6, 1985, pp. 787–88, 813. Smith quotes from a February 5, 1975, interview with the daughter of a slave, Annie Neeley, from Liberty County, Georgia: "Big husky-male slaves were encouraged to have children by female slaves" (Smith, *Slavery and Rice Culture,* 103).

20. Ryer Emmanuel, interview by A. Davis, December 16, 1937, in Rawick, ed., *The American Slave,* Vol. II, Pt. 2, p. 23.

21. Josephine Bacchus, interview by A. Davis, January 4, 1938, in Rawick, ed., *The American Slave,* Vol. II, Pt. 1, p. 22; [Ball,] *Slavery in the United States,* 72.

22. Hector Godbold, interview by A. Davis, June 28, 1937, in Rawick, ed., *The American Slave,* Vol. II, Pt. 2, p. 145.

23. A. Waiters interview, June 6, 1985, p. 788.

24. *Ibid.,* June 5, 1985, p. 726; Emmanuel interview, in Rawick, ed., *The American Slave,* Vol. II, Pt. 2, p. 13.

25. *Ibid.,* 11–12.

26. *Ibid.,* p. 14.

27. A. Waiters interview, June 5, 1985, p. 722.

28. Cannon interview, in Rawick, ed., *The American Slave,* Vol. II, Pt. 1,

pp. 189–90. Ball substantiated Mrs. Cannon's statement that male slaves wore only a long shirt and that the slaves' clothes were completely worn out. He described slaves who were working with him on his first day in a South Carolina field east of Columbia: "The old man . . . had no clothes on him except the remains of an old shirt, which hung in tatters from his neck and arms. . . . The woman who was the mother of the children, wore the remains of a tow linen shift, the front part of which was entirely gone; but a piece of old cotton bagging tied round her loins, served the purposes of an apron. The younger of the two boys was entirely naked" ([Ball,] *Slavery in the United States,* 108–109).

29. A. Waiters interview, June 6, 1985, pp. 789–90.

30. *Ibid.,* February 20, 1982, p. 98, June 6, 1985, p. 845; Emmanuel interview, in Rawick, ed., *The American Slave,* Vol. II, Pt. 2, pp. 15–16.

31. Elaine Nichols, ed., *The Last Miles of the Way: African-American Homegoing Traditions, 1890–Present* (Columbia, S.C., 1989), 5–6. John W. Blassingame wrote: "Because of labor requirements on the plantation, a deceased slave was often buried at night with the rites being held weeks later" (John W. Blassingame, *The Slave Community: Plantation Life in the Antebellum South* [New York, 1972], 33). Eugene D. Genovese gave several possible reasons for night funerals: masters refused to allow slaves time for funerals during the day; friends from neighboring plantations could attend if the funerals were at night; night funerals could have been an African pattern (Eugene D. Genovese, *Roll, Jordan, Roll: The World the Slaves Made* [New York, 1974], 197).

32. Emmanuel interview, in Rawick, ed., *The American Slave,* Vol. II, Pt. 2, pp. 24–25.

33. Grant interview, *ibid.,* 171, 173.

34. [Ball,] *Slavery in the United States,* 117.

35. Grant interview, in Rawick, ed., *The American Slave,* Vol. II, Pt. 2, p. 175.

36. Emmanuel interview, *ibid.,* p. 15.

37. A. Waiters interview, June 5, 1985, p. 725; Washington Dozier, interview by A. Davis, June 23, 1937, in Rawick, ed., *The American Slave,* Vol. II, Pt. 1, p. 331. Dozier lived on Wiles Gregg's farm, southwest of Mars Bluff. He was ninety years old when he was interviewed in 1937.

38. Sara Brown, interview by A. Davis, July 8, 1937, *ibid.,* p. 139. Mrs. Brown lived southwest of Mars Bluff near Effingham.

39. Louisa Gause, interview by A. Davis, December 2, 1937, in Rawick, ed., *The American Slave,* Vol. II, Pt. 2, p. 111. F. G. Cassidy and R. B. Le Page wrote that Cudjoe is the male day-name for Monday. A day-name is "the name formerly given (following African custom) to negro children according to their sex and the day of the week on which they were born. The system is or was used where Ashanti slaves were taken in the Caribbean, with local variations. The more common term today among the Jamaican folk is born-day name.

The names themselves survive only in pejorative senses" (F. G. Cassidy and R. B. Le Page, eds., *Dictionary of Jamaican English* [Cambridge, Mass., 1967], 144).

CHAPTER 3

1. A. Waiters interviews, October 15, 1976, pp. T49–T50, November, 1975, p. T102, February 20, 1982, p. 101. (Page numbers preceded by the letter *T* refer to the first 105 pages of one notebook of interviews, which are typed; after page T105, numbering begins again on handwritten pages.) Gregg lived within five miles of the Florence prison camp. King wrote that a party from Sherman's army was sent to Florence; however, it was turned back before reaching the camp. The main body of Sherman's army, in the march from Columbia to North Carolina, passed through Cheraw, about forty miles north of Florence (King, *Rise Up So Early,* 57). It may have been foragers from that march that Gregg saw.

2. Peter H. Wood and Karen C. C. Dalton, *Winslow Homer's Images of Blacks: The Civil War and Reconstruction Years* (Austin, 1988), 55. I am indebted to Wood for making me aware of Amy Spain. Leon F. Litwack reported similar incidents illustrating that it was unwise for slaves to express their reactions to the passage of northern troops (Leon F. Litwack, *Been in the Storm So Long* [New York, 1980], 172–75).

3. A. Waiters interview, June 5, 1985, pp. 739–40. Gregg's reaction to freedom seems to have been a very common one. Writing about the massive population movement of African Americans after the war, Eric Foner stated: "In fact, a majority of freedmen did not abandon their home plantations in 1865, and those who did generally traveled only a few miles" (Eric Foner, *Reconstruction: America's Unfinished Revolution, 1863–1877* [New York, 1988], 81).

4. A. Waiters interview, December, 1977, p. 17. John Glover, a contemporary of Gregg in Florence County, said: "Slaves didn' know what to do de first year after freedom. . . . What de slaves gwine buy land wid den, Captain? Won' a God thing to eat in dat time" (John Glover, interview by H. Davis and L. Young, prepared by A. Davis, June 28, 1937, in Rawick, ed., *The American Slave,* Vol. II, Pt. 2, p. 141).

5. A. Waiters interview, June 5, 1985, p. 742. When Waiters described where his grandfather had first moved, it seemed that he might be speaking of the area where the Gibbs family had lived, so I asked if he had ever heard his grandfather mention the Gibbses.

> I hear him say something about the Gibbs—but—told me something. I forget what he say. They give somebody an acre of land—what they give. An acre of land and a mule. The butler what work around

> the house, they give him an acre of land and a mule and tell him he can work for him and he going to pay him to work and give him an acre of land to plant his garden, to grow little things on. Time he give him an acre of land, he live about three months and he died, him and his wife.
>
> Given him a mule too. . . . He didn't have no stock to feed; his eat right along with his [Gibbs's] mule.
>
> (A. Waiters interview, July 25, 1984; pp. 330–31)

I am fairly certain that no deed was written for this "gift," and that at the freedman's death, there was no question but that the land and mule still belonged to the European American.

6. A. Waiters interviews, June 26, 1984, pp. 200–202, June 6, 1985, p. 837, February 26, 1986, p. 933.

7. *Ibid.*, June 5, 1985, pp. 746–47.

8. Catherine and Archie Waiters compiled this list of names (author's notebooks of interviews, 451).

9. A. Waiters interviews, June 26, 1984, p. 202, June 5, 1985, p. 747. Walter Gregg, son of J. Eli Gregg, mentioned Emma Gregg in a letter he wrote in 1867: "On last night I was aroused by Amy coming to the house for something for Emma who had turned in to get through her troubles. She was very cool about it as she attended to tea as usual and finished the other job before eleven o'clock" (Walter Gregg, letter to Anna Parker Gregg, October 28, 1867, photocopy in possession of Anna S. Sherman). The phrases "to get through her troubles" and "finished the other job" were euphemisms referring to childbirth; the phrase "birth of a baby" was taboo. Walter Gregg came home after the war and worked with his father, so both men are referred to as owners of the land on which Emma and Alex Gregg lived.

10. A. Waiters interviews, June 5, 1985, pp. 730–33, June 6, 1985, p. 837, March 9, 1990, n.p.; W. Gregg, letter to A. P. Gregg, December 26, 1867, photocopy in possession of Anna S. Sherman.

11. A. Waiters interview, February 26, 1986, p. 934. It is possible that this matriarchal claim on all the family's wages was an Africanism.

12. *Ibid.*, December 28, 1977, p. 14.

13. *Ibid.*, July 19, 1984, p. 226. The man in the store who gave them work was J. Wilds Wallace, son-in-law of Walter Gregg. There is no clear line of demarcation between succeeding generations in ownership of the land and the store. In Waiters' telling of his grandfather's story, the generations—J. Eli Gregg, Walter Gregg, and Charles E. Gregg and J. Wilds Wallace—all fuse together, so that any of them might be referred to as the landowner.

14. The house was one of those built about 1836 near J. Eli Gregg's home. Late in the nineteenth century, they were moved from the street to scattered locations.

CHAPTER 4

1. A. Waiters interview. This is the only quotation from Waiters that I cannot now locate in my notebooks of interviews, though I still believe I copied it from there.

2. A. Waiters interview, June 6, 1985, p. 833. Waiters used "lee little" to mean "very small." See Appendix A.

3. A. Waiters interview, June 6, 1985, p. 832. The school that Waiters recalled attending was in the lodge building on the Mt. Zion Church grounds. One teacher taught all grades. The school opened when cotton picking was completed which could be as late as January. The lodge building burned in the early 1920s.

4. A. Waiters interviews, December 29, 1977, p. 25, June 5, 1985, p. 720; [Ball,] *Slavery in the United States,* 192. Although Ball lived about sixty miles from Mars Bluff, he is quoted here because of his apt way of getting to the heart of African American life.

5. A. Waiters interviews, October 15, 1976, p. T11, September 17, 1980, p. 60, June 6, 1985, p. 783.

6. *Ibid.*, December 29, 1977, p. 41, September 17, 1980, p. 80, July 23, 1984, p. 308.

7. *Ibid.*, December 29, 1977, p. 11, June 5, 1985, p. 732. In every instance in which Waiters told me the amount of money he made on various jobs, it was only in response to my direct question. His spontaneous accounts of his jobs were centered on the work he did, never on the pay.

8. Married men with families were given priority in working through the winter on the farm; single men might not find regular farm work in the winter.

9. A. Waiters interviews, December 29, 1977, pp. 21–22, September 17, 1980, p. 68.

10. *Ibid.*, December 29, 1977, p. 40, November, 1975, p. 339.

11. Catherine Waiters, interviews, December 29, 1977, pp. 36–38, July 31, 1987, p. 1141, March 9, 1990, n.p., July 23, 1984, pp. 318–20. Mrs. Waiters traced one line of her family back to Prince Garland, who had "hair like a white person," and another line, the Brockingtons, back to Dr. William Johnson's place. Mrs. Waiters had a story from long ago about her great-great-uncle Oughta Brockington, who lived at the bottom of Dr. Johnson's hill on the edge of the Pee Dee Swamp. He refused to leave his home when warned of an approaching freshet and had to spend the night in the loft to keep from drowning.

12. *Ibid.*, December 29, 1977, p. 18, June 12, 1985, pp. 886–87, March 9, 1990, n.p.

13. *Ibid.*, December 29, 1977, p. 38. Nichols wrote of an African belief that evil spirits travel in straight lines and are distracted from doing harm by irregularly arranged writing or patterns. Quoting Maude S. Wahlman, Nichols wrote: "In the South, newsprint was placed on the walls of Southern homes and

into shoes as well, to protect against the elements or evil enslaving spirits, in the belief that 'evil spirits would have to stop and read the works of each chopped up column' before they could do any harm" (Nichols, ed., *The Last Miles of the Way,* 16).

14. A. Waiters interviews, December 31, 1977, p. 55, March 1, 1986, p. 954.

15. *Ibid.,* March 1, 1986, pp. 953–55.

16. *Ibid.,* December 29, 1977, p. 42, March 1, 1986, p. 956.

17. C. Waiters interviews, December 29, 1977, p. 39, June 12, 1985, p. 908. African Americans' yards were swept daily; consequently, they were hard-packed dirt with no grass anywhere. See Appendix A.

18. A. Waiters interview, December 29, 1977, p. 42. There were a number of lodges at Mars Bluff. Catherine Waiters said that her ancestors, who lived at the north edge of the settlement, went to lodge meetings at Sand Bed Church, later called Mt. Pisgah Church. Three lodges met there: Joint Stock, Ruth Eye, and Rising Star (C. Waiters interview, March 9, 1990, n.p.). Hester Waiters, who lived at the south side of Mars Bluff, said that three lodges met on the Hart place, where she had lived as a child: Odd Fellow, Joint Stock, and St. Joseph (Hester Waiters, interview, August 14, 1987, p. 1234). Stressing the importance of burial benefits, Nichols quoted from Jean Masamba and Richard Kalish: "to die without proper ceremonies . . . threatens a forgotten afterlife and the possibility of a distant relationship with God" (Nichols, ed., *The Last Miles of the Way,* 23–24).

19. C. Waiters interview, June 12, 1985, pp. 910–12. Eleven of Mrs. Waiters' deliveries were attended by midwives. The twelfth child was born in the hospital because she had elevated blood pressure.

20. *Ibid.,* February 19, 1986, pp. 926–28.

21. Archie and Catherine Waiters, interview, February 4, 1986, p. 921. Wood wrote: "The West African and Carolina climates were similar enough so that even where flora and fauna were not literally transplanted, a great deal of knowledge proved transferable. African cultures placed a high priority on their extensive pharmacopoeia, and since details were known through oral tradition, they were readily transported to the New World." John Brickell, who wrote a natural history of North Carolina in 1737, said about slaves there: "On Sundays, they gather Snake-Root, otherwise it would be excessive dear if Christians were to gather it" (Brickell, quoted in Wood, *Black Majority,* 120–21).

22. A. and C. Waiters interview, February 4, 1986, pp. 921, 925. Waiters said there were two different kinds of snakeroot. The kind Mrs. Waiters gathered was black snakeroot; the other kind was "saience" (Sampson) snakeroot. "It got a longer leaf than this here one got. . . . They grow four leaves in the place of three. This here [black snakeroot] don't grow but three leaves on it. . . . They all work the same way, just alike." According to the information Zora Hurston gathered: "a. Sampson snake root is powerful because it has a root with

four prongs and there are four corners to the world. Tap root grows straight down, and that points to the center of the world. b. Black snake root is powerful because it is like a bunch of threads. Therefore it will bind together hearts and prosperity and will tangle up and hinder those you don't want to succeed" (Zora [Neale] Hurston, "Hoodoo in America," *Journal of American Folk-Lore,* XLIV [October–December, 1931], 396). Josephine Bacchus, a contemporary of Waiters' grandfather in Marion County, confirmed Waiters' use of snakeroot as an appetite stimulant: "Dey was black snake root en Sampson snake root. Say, if a person never had a good appetite, dey would boil some of dat stuff en mix it wid a little whiskey en rock candy en dat would sho give dem a sharp appetite" (Bacchus interview, in Rawick, ed., *The American Slave,* Vol. II, Pt. 1, p. 24).

23. C. Waiters interview, June 12, 1985, pp. 892–93; A. and C. Waiters interview, February 4, 1986, p. 921. When Lawrence Swails, a botanist, and I visited the area where Mrs. Waiters found the "quincy-like root," I tentatively pointed out the plant Mrs. Waiters was referring to. Swails identified that plant as *Vaccinium arboreum* (sparkleberry) (Lawrence Swails, conversation with author, Mars Bluff, S.C., June 29, 1992. Notes from most conversations cited in the notes are filed alphabetically in conversation volume of interview notebooks, to be deposited in the Caroliniana Library, University of South Carolina, Columbia, S.C.).

24. A. Waiters interview, February 4, 1986, p. 923. Waiters pronounced the word as if it were spelled "amry." He described the plant as "up tall as this banister and then got a long leaf like a weeping willow and leaf droop down. . . . Then three prongs will be in the top of it. That one lay over there. That one lay over there. That one lay that a way." Vennie Deas-Moore listed amyroot in a chart entitled "Medically Valid Sea Islands Pharmacopoeia" and described amyroot as "roots steeped in whiskey . . . used to treat colds and fever" (Vennie Deas-Moore, "Home Remedies, Herb Doctors, and Granny Midwives," *World and I* [January, 1987], 481).

25. C. Waiters interviews, December 31, 1977, p. 51, June 12, 1985, p. 890. These quilts were probably similar in design to those made by African Americans at Mars Bluff one hundred years earlier. George W. McDaniel described quilts that slaves made in Maryland. They consisted "of large blocks of [heavy] cloth laid out in rectangular patterns across the backing" (George W. McDaniel, *Hearth and Home: Preserving a People's Culture* [Philadelphia, 1982], 109).

26. C. Waiters interviews, June 12, 1985, pp. 899–900, December 29, 1989, pp. 19–20, 38–39, 51.

27. While Mrs. Waiters did the cooking at home, her husband sometimes helped people prepare for large gatherings, cooking barbecue, fish stew, or chicken bog (A. Waiters interview, December 31, 1977, pp. 51–53).

28. *Ibid.,* November 15, 1976, pp. T91–T92, December 29, 1977, pp. 38, 51, June 12, 1985, pp. 887, 904, November 15, 1975, p. T88.

29. *Ibid.,* June 12, 1985, pp. 895–96.

30. A. and C. Waiters interview, December 29, 1977, pp. 35–36. Melville J. Herskovits wrote of the Trinidad shouters singing improvisations dictated by "the spirit" on the subject of the trumpet waking the dead. See Melville J. Herskovits, *The New World Negro: Selected Papers in Afroamerican Studies,* ed. Frances S. Herskovits (Bloomington, Ind., 1966), 346.

31. C. Waiters interviews, December 29, 1977, p. 36, June 12, 1985, p. 914.

32. A. Waiters interview, September 17, 1980, pp. 80–81.

33. *Ibid.,* March 1, 1986, pp. 944, 946–48.

34. *Ibid.,* June 12, 1985, pp. 797–802, March 1, 1986, pp. 944–46, 948–49, 951, 952.

35. *Ibid.,* June 6, 1985, p. 889. Tom Waiters, Archie's brother, had arranged to buy this lot, but he had paid only $10.00 on it. He agreed that they could buy it instead, and the owner, J. W. Wallace, sold it to them for $150.00. The Waiterses said that they had one other opportunity to buy land, also in the 1950s. A man who was moving away offered to sell his small acreage on the highway halfway between Mars Bluff and Florence for $600.00. The Waiterses wanted to buy it in cooperation with their son and daughter-in-law, but they did not have enough money (A. and C. Waiters interview, March 9, 1990, n.p.).

36. C. Waiters interviews, March 23, 1984, p. 127, June 12, 1985, pp. 889, 899; Annie Lee Waiters Robinson, interview, Mars Bluff, S.C., March 23, 1984, p. 127. This account of land ownership and access to schools calls for an explanation of why there has been no mention of politics. The reason is that Archie Waiters was not apt to mention politics or voting. Silence on those subjects may have been a survival skill learned during Reconstruction and still useful one hundred years later. The only time Waiters mentioned voting was when he recalled that for the first two years that he had access to the polls, he would not vote because he was told that his vote would not be counted. As Waiters put it, "They say Black people could vote but the vote wouldn't be no good." He added, "Now I go and vote" (A. Waiters interview, June 6, 1985, p. 842).

37. A. Waiters interviews, June 6, 1985, pp. 797–98, March 1, 1986, p. 951. Catherine Waiters gave her husband's years of employment at Koppers Industries as 1947 to 1969 (C. Waiters, letter to author, May 28, 1990). A local historian wrote: "On June 30, 1954, Koppers Company, Incorporated, bought approximately 145 acres of land . . . just beyond the eastern limits of the city of Florence. . . . It constructed a large wood preserving plant for treating poles, piling, cross ties, and fence posts" (Henry E. Davis, "A History of Florence, City and County, and of Portions of the Pee Dee Valley, South Carolina" [Typescript, 1965, in Special Collections, Francis Marion College Library, Florence, S.C.], Chap. 18, p. 15. Numbering of pages in each chapter begins with page one.) Koppers' public relations department reconciled these statements by noting that a plant was established near Florence in 1946 and that Koppers had taken it over in 1954.

INTRODUCTION TO PART TWO

1. Lucas Dargan, conversation with author, Mars Bluff, S.C., *ca.* 1970.

CHAPTER 5

1. Many people picture a swamp as land covered by water. The way the word *swamp* is used at Mars Bluff conforms to Webster's definition: A swamp is wet, spongy land saturated and sometimes partially or intermittently covered with water.

2. The preceding dialogue was reconstructed from the following interviews with Waiters: October 15, 1976, pp. T38–T49, December 31, 1977, pp. 47–49, September 17, 1980, p. 85, February 20, 1982, pp. 92–97, July 25, 1984, pp. 322–25, July 31, 1987, pp. 1109, 1143. When asked if he could remember anyone paying Willie Scott for rice, Waiters explained: "No, he give rice. He never charged nobody for rice. They work on his crop—break corn, pick cotton. He give you whole 200 pound, one sack. Then you share yours with whoever you want to because you ain't going to eat it all" (February 20, 1982, p. 96).

3. *Ibid.,* October 15, 1976, p. T43.

4. *Ibid.,* February 20, 1982, p. 91, March 23, 1984, pp. 112–20, July 31, 1987, p. 1147.

5. A. Waiters interviews, September 17, 1980, p. 85, March 23, 1984, p. 114, June 5, 1985, pp. 756–58, 762, 767, 770–71, July 31, 1987, pp. 1101, 1109–10, 1115, 1140, 1145, 1147, June 28, 1988, pp. 1309, 1311, 1313; Mattie Smalls Gregg, interview, August 3, 1987, p. 1176.

6. A. Waiters interviews, October 15, 1976, pp. T38, T48, September 17, 1980, p. 85, February 5, 1982, p. 91, March 23, 1984, pp. 120–21, June 26, 1984, pp. 203–206, July 19, 1984, pp. 250–51.

CHAPTER 6

1. M. R. Werner, *Julius Rosenwald: The Life of a Practical Humanitarian* (New York, 1939), 127–33 (quotations on 130); Edwin R. Embree and Julia Waxman, *Investment in People: The Story of the Julius Rosenwald Fund* (New York, 1949), 40–51.

2. One woman had heard her father state a dollar amount that each family was supposed to pay.

3. Matthew Williamson, interview, May 14, 1985, p. 666. Stroyer, in his book about his life in Sumter County a generation earlier, spoke of this custom:

"It was a custom among the slaves not to allow their children under certain ages to enter into conversation with them" (Stroyer, *My Life in the South,* 23). Lizzie Davis, of Marion County, in recalling what she had heard her father say, concluded with "Cose I speak bout what I catch cause de olden people never didn' allow dey chillun to set en hear dem talk no time. No, mam, de olden people was mighty careful of de words dey let slip dey lips" (L. Davis interview, in Rawick, ed., *The American Slave,* Vol. II, Pt. 1, p. 293).

4. Claudia Williamson Williams, interview, July 17, 1984, pp. 367–68. Mrs. Williams recalled going away to school: "I lived with a widow, white lady. She wanted somebody to stay with her. I'd get up in the morning and get breakfast and do a little chores around the house, and I'd get dressed and go on to Claflin to school, and that's the way I got my education. . . . I finished the L.I. [licensed to instruct] degree there. . . . It took me longer to get my college degree because I would work some and then go to summer school. It was backwards and forwards like that" (*ibid.*).

5. Holt wrote that A. H. Howard of Marion County was a representative in the Reconstruction legislature of 1874 (Thomas Holt, *Black over White* [Urbana, Ill., 1977], 234). "A. H." were the initials Tony Howard used on deeds.

6. C. Williams interview, July 17, 1984, pp. 377–81, 390–91.

7. Williamson interview, May 14, 1985, pp. 654–58, 670.

8. C. Williams interview, July 17, 1984, pp. 372, 388; Williamson interview, May 14, 1985, pp. 681, 690.

9. Mable Smalls Sellers, interview, February 27, 1984, pp. 128–33. Mattie Gregg said that Lewis Smalls lived from 1880 to 1935; his entire life was spent in the Mars Bluff area (Gregg interview, August 3, 1987, pp. 1171–72).

10. Gregg interview, August 3, 1987, pp. 1177–82. From the descriptions given by Archie Waiters of the location of Jim's rice field and by Mattie Gregg of Alfred Robinson's field, I gather that the fields were adjacent or that the men used the same field.

11. Frances Johnson, interview, March 12, 1984, pp. 185a, 185k. Miss Johnson's father had bought the farm about 1935 or 1937, and began planting rice there at that time. Before moving there, he had lived and planted rice on a European American's land east of Liberty Church, just a few miles from the farm he bought.

12. *Ibid.,* 185a–185c.

13. *Ibid.,* March 12, 1984, p. 187, July 2, 1984, pp. 193–94, 197.

14. *Ibid.,* March 12, 1984, pp. 185c–185e, 190, July 2, 1984, pp. 194–96.

15. *Ibid.,* March 12, 1984, pp. 185h–185j. After husking the rice using a mortar and pestle, Miss Johnson said she would "put it in a dishpan and fan all that trash out it. Then you got to put it back and you get it good and clear, then you get some shucks. . . . Corn shuck. . . . and tear them up, in little strips and throw it in the rice, in the mortar. And then take that pestle and go back at it

again. . . . to clear the rice. . . . Make it pretty. . . . Make it white. I used to do that more times than a little" (March 12, 1984, pp. 185h–185i).

16. *Ibid.*, March 12, 1984, pp. 185a, 185l, 186, 190, July 2, 1984, pp. 191–93.

17. Leon H. Coker, interview, March 14, 1984, pp. 143–45.

18. A 1943 government bulletin on upland rice revealed the U.S. Department of Agriculture's interest in noncommercial forms of rice production. This attention to rice cultivation stemmed from food shortages caused first by the depression and then by World War II. The bulletin defined upland rice as that grown on land not irrigated or submerged during the growing season. It reported that 2,600 acres of upland rice were grown in Alabama, Florida, Georgia, Mississippi, and South Carolina in 1939, with an average of 2.3 acres per farm. "In the 'live-at-home program' in the Southern States, rice fits in well for people who are accustomed to using it in their regular diet. The crop can be grown on a small scale on land less well suited for other crops by families who can care for the crop when not occupied with other farm work. . . . On many farms in the south Atlantic and Gulf states there are low moist areas that often are too wet for growing most crops, but on which upland rice is and can be grown to advantage" (Jenkin W. Jones, "Upland Rice" [Washington, D.C., 1943[, multigraphed, 1–2). Steven Taylor's son said that his father, a European American, began planting rice at Mars Bluff in the 1930s and continued into the 1940s. He used a spot of "dark, moist-type soil" in a dry field. Taylor was a friend of the county agricultural extension agent and may have been encouraged by him to plant rice to promote self-sufficiency. Taylor stopped growing rice when the rice-husking mills closed in the early forties (Harmon Taylor, interview, July 31, 1987, pp. 1155–56).

CHAPTER 7

1. Ida Ellison Zanders, interviews, August 4, 1987, pp. 1190–91, 1196, 1202–1203.

2. *Ibid.*, August 7, 1987, p. 1206, June 30, 1988, n.p. Fanny Ellison's house and garden were immediately west of Rogers Library at Francis Marion College. The house was southwest of the library entrance; the garden and rice were northwest of the entrance. These relative positions are known because the house was not moved until after the library was built.

3. Judith Ann Carney, "The Social History of Gambian Rice Production: An Analysis of Food Security Strategies" (Ph.D. dissertation, University of California, Berkeley, 1986), 72.

4. C. Waiters interview, June 6, 1988, pp. 1315–16. Mrs. Waiters said that Fanny Ellison had left Mars Bluff by 1926, perhaps several years earlier. Atleene

Pinkney wrote that the year of Fanny Ellison's death was 1943 (Atleene Pinkney, letter to author, March, 1988).

Chapter 8

1. Joseph Broadwell, conversation with author, Mars Bluff, S.C., May 12, 1984.

2. H. Waiters interview, August 14, 1987, pp. 1227–30.

3. *Ibid.*, 1219–26. Hester Waiters was born in 1904 and was eleven or twelve years old the year they planted rice.

4. Willie H. Bailey, telephone conversation with author, August 23, 1987, and interview, July 7, 1988, n.p. Benny Bailey's full name was Christopher Benny Bailey.

5. Ruth Lee Scott Bailey, conversation with author, Mars Bluff, S.C., July 7, 1988.

Chapter 9

1. Lacy Rankin Harwell, conversation with author, Mars Bluff, S.C., n.d. Mrs. Harwell was the only European American that I ever heard say she knew of an African American growing rice at Mars Bluff. However, I began my search for rice growers in the area in the 1970s. I am confident that if I had started twenty years earlier I would have found more European American landowners who knew about the African American rice growers.

2. Isabell Daniels Smith, telephone conversation with author, March 13, 1984. Mrs. Smith was born in 1909. She could not go with me to show me the field. Later, when I searched the area she had described, I could not be sure exactly where the field was.

3. Mary Daniels Washington, interview with author, Mars Bluff, S.C., February 23, 1984, p. 419. Mrs. Washington recalled that after her father had threshed the rice, he took it to the mill to "clean it." He stopped planting rice when he moved. That date is placed in the 1920s, based on Mrs. Washington's information: they moved when she was a small child and she was "way past 65" in 1984.

4. Lacy Rankin Harwell, conversation with author, Mars Bluff, S.C., July 16, 1984. Mrs. Harwell identified two of the early settlers as Bostick and Pearce.

5. According to James Harwell, after rice production ceased, "the fields were used . . . for corn for a while, and then pasture for a while, and then [were] reclaimed by nature" (James R. Harwell, letter to author, June, 1984).

6. This was a floodgate on the east side of the rice fields. Another floodgate

was recalled by Catherine Willis. As a child she fished at a floodgate on the south side of the old rice fields near the causeway that Mrs. Harwell described as immediately behind her home (Catherine Willis, telephone conversation with author, May 17, 1992).

7. Deeds show that in addition to "Bostick and Pearce," whom Mrs. Harwell had named, John Wright, John Flemor, and Micajah Pippin were early settlers on what could have been the rice field land. None of the documents examined refers to rice except Tristram Bostick's estate inventory for 1819, which lists cotton, oats, cattle, sheep, hogs—and two empty rice casks (Tristram Bostick, Appraisement of Estate, January 14, 1820, at Marion County Court House, Marion, S.C.). Two casks suggest subsistence rice cultivation or the purchase of rice rather than commercial rice production. James Lane was a likely person to have owned part of the commercial rice fields, but his estate records for 1844 contain no clue (James Lane, Appraisement of Estate, October 7, 1844, at Marion County Court House, Marion, S.C.).

8. Deed from James Lane to Dr. William R. Johnson for 1,631 acres, May 20, 1844, Marion County, S.C., Deed Book S, 399–400, in South Carolina Department of Archives and History, Columbia, S.C.

9. McCarthy, *Footprints,* 74. Dr. Johnson was born in Brittons Neck, Marion County, in 1813. He graduated from medical college in Charleston in 1838, settled in Mars Bluff as a young man, and served several terms in the South Carolina legislature. I am indebted to Ann Niemeyer Williamson for information about Dr. Johnson (Ann Niemeyer Williamson, letter to author, May 22, 1992; Medical University of South Carolina, letter to Dr. William H. Johnson, November 10, 1981, photocopy in possession of Ann Niemeyer Williamson, Darlington, S.C.; W. W. Sellers, *A History of Marion County, South Carolina* [Columbia, S.C., 1902], 567, 568).

10. The article is DeLancey Evans, "Re-establishment of the Rice Industry in the United States After the Civil War," *Rice Journal,* XXIII (September, 1920), 27–28, 36–39. This article has not a single word about the rice fields on the Harwell farm.

11. *U.S. Census: Original Agriculture, Industry, Social Statistics, and Mortality Schedules for South Carolina, 1850,* Schedule 4, Productions of Agriculture in Marion District, South Carolina, South Carolina Archives Microcopy No. 2, pp. 217–18, in South Carolina Department of Archives and History, Columbia, S.C. I am indebted to Marion C. Chandler for this record. I am also grateful to Horace Rudisill for informing me of the existence of pertinent agricultural records.

12. Horace Fraser Rudisill, letter to author, May 18, 1988. I am indebted to Rudisill for providing a typed copy of the easement. It read:

> I, William T. Wilson, of the district and State aforesaid . . . in consideration of the benefits and advantages which I shall derive in the

> way of drainage . . . have granted to William R. Johnson the right privilege and easement forever to open a ditch and to continue open and drain over and through the same . . . of the following width and depth to wit the said ditch to be opened eight feet wide and two feet deep or deeper if said Johnson shall find it to his interest to cut it deeper . . . through and over lands of which I the said Wilson am at present seized and possessed . . . lying on the South side of the Great Pee Dee River and on the Back Swamp near the Duck Ponds . . . said ditch to commence in the run of Back Swamp on the lines dividing said Johnson's land from mine then to run a Northeast course down the run of Back Swamp to the line which divides my land from the lands of Eli M. Bostick. . . . [signed] William T. Wilson, Witness: Samuel N. Lane, Thomas M. Lane.
>
> (William T. Wilson to William R. Johnson, easement for ditch, August 3, 1846, Marion County, S.C., Deed Book T, 375–76, in South Carolina Department of Archives and History, Columbia, S.C.)

At the same time, Dr. Johnson got a similar easement from Eli M. Bostick, and four years later Dr. Johnson and Wilson got a similar easement from John A. Brown (Eli M. Bostick to William R. Johnson, easement for ditch, August 3, 1846, Marion County, S.C., Deed Book T, 376–77, and John A. Brown to W. R. Johnson and W. T. Wilson, easement for ditch, 1850, Marion County, S.C., Deed Book W, 386–87, both in South Carolina Department of Archives and History, Columbia, S.C.).

13. Grant interview, in Rawick, ed., *The American Slave,* Vol. II, Pt. 2, pp. 171, 173.

14. W. R. Johnson, 1844 plat, Marion County, S.C., Plat Book B, 236, in South Carolina Department of Archives and History, Columbia, S.C. The fact that Dr. Johnson began building his house about 1850 on land immediately south of the old rice fields would suggest that he acquired a part of the rice field area in the 1844 deed. There are no other records of deeds to him in that area until a deed from the Pearce family in 1859 (Robert H. Pearce *et al.* to Dr. William R. Johnson, deed for 1,744 acres, December 28, 1859, Marion County, S.C., Deed Book Y, 523–24, in South Carolina Department of Archives and History, Columbia, S.C.).

15. Will of Hannah Thompson, October 21, 1856, at Marion County Court House, Marion, S.C.

16. The flimsy clues possibly associating Samuel Gibson with the Harwell fields are as follows: According to the deed from Lane to Johnson, Robert Napier owned land adjacent to Dr. Johnson. According to the easement from Bostick to Johnson, Eli M. Bostick owned land near Johnson's land. A plat of Rob-

ert Napier's land indicates that it was bounded by lands of E. M. Bostick and Samuel Gibson (Robert Napier, plat, November 3, 1846, Marion County, S.C., Plat Book B, 259, in South Carolina Department of Archives and History, Columbia, S.C.). Could Samuel Gibson have been farming in the area of the Harwell farm rice fields? I doubt it. Gibson land was generally closer to the Great Pee Dee River.

17. Timothy G. Dargan, "Canals Leading to Irrigated Rice Fields at Mars Bluff," map drawn for author, July 11, 1988, in author's conversation notebook. The canals were located on a United States government aerial photograph (Washington, D.C., 1975). I am indebted to Dargan for this contribution to our understanding of the fields.

18. A. Waiters interview, July 25, 1984, pp. 327–28. At Mars Bluff, along the Great Pee Dee River and Jefferies Creek, the remains of embankments are common and seem to suggest old rice fields. But these embankments usually were built to keep water out of fields where dryland crops were cultivated. De Bow wrote that the swamplands along the rivers above the tidewater area, "having been reclaimed, are protected by dams (*levees*) from the destructive influx of heavy freshets to which those rivers are annually subject. Leaving out of view, for the present, the tide-swamp, the cotton plantation nearest the seacoast which has been dammed on the Pee Dee, is Mr. Gibson's, near Marr's Bluff, where the Manchester Railroad is to cross" (J. D. B. De Bow, "Seacoast Crops of the South," *De Bow's Review,* III [June, 1854], 603–604). Harry Harllee, a landowner in the Pee Dee Swamp area, said miles of dikes were built there for flood control (Harry Harllee, telephone conversation with author, Mars Bluff, S.C., May 12, 1984). In the southwest part of Mars Bluff, Joe Broadwell showed me the levee that ran along Jefferies Creek on his land. Broadwell said that it was used to hold back creek water when the land beside the creek was used by the early settlers. Alex Gregg, who was raised on a farm on Jefferies Creek east of the Broadwell farm, recalled a levee there (A. Waiters interview, July 25, 1984, p. 333).

CHAPTER 10

1. Edward McCrady, *The History of South Carolina Under the Proprietary Government, 1670–1719* (New York, 1897), 125–26; Wallace, *South Carolina: A Short History,* 32; A. S. Salley, *The Introduction of Rice Culture into South Carolina,* in Bulletin of the Historical Commission of South Carolina No. 6 (Columbia, S.C., 1919), 3.

2. James M. Clifton, "The Rice Industry in Colonial America," *Agricultural History,* LV (July, 1981), 269–70.

3. A. S. Salley, "The True Story of How Madagascar Gold Seed Was In-

troduced into South Carolina," in *Contributions from the Charleston Museum,* ed. E. Milby Burton, VIII (1936), 51; Duncan Clinch Heyward, *Seed from Madagascar* (Chapel Hill, 1937), 4–5. Note that this story deals only with the introduction of Madagascar gold rice. Lewis Cecil Gray makes clear that other varieties of rice were present in the Carolina colony (Lewis Cecil Gray, *History of Agriculture in the Southern United States to 1860* [Washington, D.C., 1933], I, 277–78). Given that Woodward received credit for the first successful cultivation of Madagascar gold, it is interesting to note that "among those that are known to have brought numbers of blacks with them from the West Indies to South Carolina in the early years [is] . . . Henry Woodward [who brought] seven" (David LeRoy Coon, "The Development of Market Agriculture in South Carolina, 1670–1785" [Ph.D. dissertation, University of Illinois, 1972], 154). One might speculate whether it was Woodward or one of the seven African Americans who deserved credit for his success in rice cultivation.

Clifton reviewed the history of conflicting stories about the introduction of rice into Carolina (Clifton, "The Rice Industry," 271–72). One widely told story about the introduction of Madagascar gold gives Landgrave Smith credit. See J. D. B. De Bow, "Rice," *De Bow's Review,* I (April, 1846), 329–30; David Ramsay, *Ramsay's History of South Carolina, from Its First Settlement in 1670 to the Year 1808* (1809; rpr. Newberry, S.C., 1858), II, 113–14. By the time Phillips repeated the story, the type of rice had been omitted, and Smith received full credit for the introduction of rice into Carolina: "Among the crops tried was rice, introduced from Madagascar by Landgrave Thomas Smith about 1694, which after some preliminary failures proved so great a success" (Phillips, *American Negro Slavery,* 87).

4. De Bow, "Rice," 330; also see Wallace, *South Carolina: A Short History,* 188. This period was called the inland swamp period of rice cultivation (Heyward, *Seed from Madagascar,* 11–12). Mark Catesby wrote of Carolina, "*Rice Land* is most valuable, though only productive of that grain, it being too wet for any thing else" (Mark Catesby, *The Natural History of Carolina, Florida, and the Bahama Islands* [1731; rpr. Savannah, Ga., 1974], 13).

5. Heyward, *Seed from Madagascar,* 7; John Archdale, "A New Description of That Fertile and Pleasant Province of Carolina," in *Narratives of Early Carolina, 1650–1708,* ed. Alexander S. Salley, Jr. (New York, 1911), 289.

6. Heyward, *Seed from Madagascar,* 14; Herbert Ravenel Sass, "The Rice Coast: Its Story and Its Meaning," in Alice Ravenel Huger Smith, *A Carolina Rice Plantation in the Fifties: Thirty Paintings in Water-Colour* (New York, 1936), 22. In addition to the reservoir system of irrigation described by Sass, some inland rice fields were irrigated by diverting water from freshwater rivers (State Board of Agriculture of South Carolina, *South Carolina Resources and Population, Institutions and Industries* [Charleston, S.C., 1883], 56).

7. Sass, "The Rice Coast," 22. The trunks were designed so that when the

outer gate was opened, fresh water from the river could flow into the rice fields; when the inner gate was opened, the fields could be drained (Heyward, *Seed from Madagascar,* 5–6, 12–13). Rogers gave early evidence to support the theory that tidal-flow irrigation evolved rather than was invented at a certain time by a specific person (Rogers, *The History of Georgetown County,* 332). Some credit McKewn Johnston, a planter in the Georgetown area, with introducing the tidal-flow system in 1758. Perhaps it is more appropriate to say that he perfected the system (Henry C. Dethloff, *The History of the American Rice Industry, 1685–1985* [College Station, Tex., 1988], 19–20). David Doar gives details about the schedule for repeated floodings (David Doar, "Rice and Rice Planting in the South Carolina Low Country," in *Contributions from the Charleston Museum,* ed. E. Milby Burton, VIII [1936], 14–15).

8. Wallace, *South Carolina: A Short History,* 189; James M. Clifton, ed., *Life and Labor on Argyle Island: Letters and Documents of a Savannah River Rice Planter, 1833–1867* (Savannah, Ga., 1978), viii.

9. Littlefield wrote of this period: "Despite the ethnocentrism that infects all national groups and expands whenever divergent cultures come into close proximity, these attitudes of derogation may not have been so common, or at least not so blinding, as they were to become when racism was elevated to the status of a cornerstone of the American way of life" (Littlefield, *Rice and Slaves,* 5).

10. Phillips, *American Negro Slavery,* 339, 342–43.

11. W. E. Burghardt Du Bois, *Black Reconstruction in America: An Essay Toward a History of the Part Which Black Folk Played in the Attempt to Reconstruct Democracy in America, 1860–1880* (New York, 1935), 719. Sass praised European Americans for developing the tidal-flow system of irrigation while "the only labor at the disposal of the settlers who accomplished the feat was of the most unskilled character, African savages fresh from the Guinea coast" (Sass, "The Rice Coast," 23).

12. Coon, "The Development of Market Agriculture," 337.

13. Wood, *Black Majority,* 56.

14. *Ibid.,* 43–44, 331, 105, 165–66, 195, 229, 275, 324–25. Wood traced the history of the development of this attitude from its beginning when African Americans first outnumbered European American settlers in the early eighteenth century to its intensification after the Stono Rebellion in 1739.

15. *Ibid.,* 26, 35, 57–58, 59–61.

16. *Ibid.,* 9, 19–25, 143, 95. As to when the first African Americans arrived in the colony, Wallace wrote: "Governor Sayle, we know, brought in three Negroes in the first fleet and a fourth in September 1670. Captain Henry Brayne of the *Carolina* owned 'a lusty negro man' in South Carolina in November, 1670. Three slaves arrived in February, 1671, and from 1671 to 1674 Lady Margaret Yeamans imported eight. The same allotment of land was allowed as for

white settlers" (Wallace, *South Carolina: A Short History,* 31). W. Robert Higgins attributed the first African Americans to "the wife of the proprietary governor, Sir John Yeamans" (W. Robert Higgins, "The Geographical Origins of Negro Slaves in Colonial South Carolina," *South Atlantic Quarterly,* LXX [Winter, 1971], 34).

17. Wood, *Black Majority,* 57, 62. The statement that Englishmen had "now found out the true way of raising and husking Rice" carries a point not appreciated in today's mechanized world. I am reminded of a comment made as recently as 1980 by a European American in Florence. She said that she had grown a little rice in the 1940s when there was a national emphasis on victory gardens and self-sufficiency, but she had only raised it one year because she could not figure out how to husk it.

18. Joyner wrote of rice culture in South Carolina: "The early technological knowledge was supplied by Africans, not Europeans. To support this statement it is not necessary to establish that all, or even most, of the Africans who came to South Carolina were experienced in rice culture. All that is necessary is to point out that none of the Europeans, whether from the British Isles, Western Europe, or the Caribbean, had any experience with rice culture at all" (Joyner, *Down by the Riverside,* 13).

19. Littlefield, *Rice and Slaves,* 86, 98.

20. *Ibid.,* 113. Mullin wrote, "They [Carolinians] preferred slaves from Senegambia (present-day Senegal and Gambia) and Bumbara and Malinke people, and they disliked short Africans. Their second choice was slaves from the Gold Coast (Ghana). Farther south in Nigeria were men like Olaudah Equiano, whom they avoided, arguing that Iboes were despondent and sometimes suicidal. While only 2.1 percent of all eighteenth-century South Carolina slaves were from this region, nearly 2 in 5 (or 40 percent) of Virginia's slaves were from Nigeria. Instead, Carolinians obtained 40 percent of their slaves from farther south, in Angola, while only 16 percent of Virginia's slaves came from that region" (Mullin, ed., *American Negro Slavery,* 48; also see Philip Curtin, *The Atlantic Slave Trade: A Census* [Madison, Wis., 1969], 157). Littlefield wrote, "The high percentage of Angolas can be assigned to a combination of acceptability and availability." Littlefield did not believe that it was due to their knowledge of rice: "Despite intimations to the contrary, rice culture does not appear to have been significant in that region [Angola]." He continued: "Although the crop is an important staple in parts of the region in the modern period, rice cultivation seems to have been introduced relatively late—with the advent of the Europeans—and, in some areas, long after colonization" (Littlefield, *Rice and Slaves,* 113, 109, 109*n*).

21. Littlefield, *Rice and Slaves,* 177.

22. Joyner, *Down by the Riverside.* The entire book portrays the merger of the two cultures.

23. Alex West, "The Strength of These Arms: Black Labor—White Rice," Video (Durham, N.C., 1988).

CHAPTER 11

1. R. Q. Mallard, *Plantation Life Before Emancipation* (Richmond, 1892), 31, 35. Writing about African Americans on the irrigated rice plantations in South Carolina, Doar stated: "The family was allowed land for a garden, and they also could plant rice, if they wished, on outside margins of the river, a privilege which a great many availed themselves of, judging by the little fields, which could be seen on the plantations" (Doar, "Rice and Rice Planting," 32). Dr. Richard Schulze of Turnbridge, an old rice plantation in the tidewater region, reported that there had been small plots of rice grown in one area on his plantation (Richard R. Schulze, letter to author, January 9, 1989).

2. William Law listed some of the crops grown by a Darlington man in 1814: 8 acres of wheat, 10 acres of oats, 6 acres of potatoes, 1 acre of slips, 3 acres of rice, 75 acres of peas, and 20 acres of fodder (William Law Book, 1813, in William Law Papers, Box 2, File 2, Manuscript Department, William R. Perkins Library, Duke University, Durham, N.C.).

3. Inventory of John Frazer [*sic*] Estate, in John Fraser Papers, in possession of Horace Fraser Rudisill, Darlington, S.C. I am indebted to Rudisill for allowing me to use files that he has spent many years collecting.

4. John Gregg, Appraisement of Estate, October 24, 1839, at Marion County Court House, Marion, S.C. The estate of James Gregg, who died at Mars Bluff in 1802, was appraised as having a rice crop one-tenth the value of the corn crop. There would have been a considerably larger difference if measured by volume (James Gregg, Appraisement of Estate, March 16, 1802, *ibid.*).

5. William Curry Harllee, *Kinfolks: A Genealogical and Biographical Record* (New Orleans, 1934), 726. The amount of rice suggests that Harllee's may have been one of the three places where Alex Gregg helped with the rice crop. This farm was only about two miles east of Gregg's boyhood home.

6. Henry L. Pinckney Plantation Book, 1850–1867, pp. 57–58, in Manuscript Department, William R. Perkins Library, Duke University, Durham, N.C.

7. In all four cases, the number of slaves was taken from the four sources previously listed regarding rice production. Pinckney noted that his work force was reduced to thirty after the Yankee raid.

8. "Traditional methods of cultivating *O. glaberrima* have been practiced in parts of the region [West Africa] for perhaps 15 centuries or longer" (Scott R. Pearson, J. Dirck Stryker, and Charles P. Humphreys, *Rice in West Africa: Policy and Economics* [Stanford, 1981], i). Both Littlefield and Olga Linares de Sapir

described rice-growing practices in Africa that were similar to practices at Mars Bluff (Littlefield, *Rice and Slaves,* 86, 91–92, 98; Olga Linares de Sapir, "Agriculture and Diola Society," in *African Food Production Systems,* ed. Peter F. M. McLoughlin [Baltimore, 1970], 206). Of course, adaptations were necessary to accommodate African knowledge to Carolina, because Africa had a tropical climate and inundating seasonal rains while Carolina had a semitropical climate and an annual rainfall of about forty-four inches.

9. Davis, "A History of Florence," Chap. 15, p. 16. I am grateful to Wayne King of Francis Marion College for making me aware of the Davis manuscript. It is a gem, filled with such fascinating details as the following: "In the early 1890's [in Davis' boyhood community in Williamsburg County], a movement was initiated to expand the growing of upland rice. . . . Crops were planted on nearly every farm, and the prospects for the revival of an industry were most promising. On our plantation we planted two acres of rice on a very rich bottom, and confidently expected a large yield. We did not make a pint on the two acres. The bill bug, an insect similar to the cotton boll weevil, destroyed the crop" (Chap. 15, p. 17). On the subject of small fields of rice in the pine belt, Drayton wrote: "Rice may be said to be solely the produce of the lower country. It is sometimes grown in the middle country; but of small quantity, more for the use of its inhabitants, than for the purposes of sale" (John Drayton, *A View of South Carolina as Respects Her Natural and Civil Concerns* [Charleston, S.C., 1802], 116). Sellers wrote that rice "was never raised to much extent in Marion County; only raised for domestic use. A few old rice plantations were in the lower part of the county, contiguous to the river" (Sellers, *A History of Marion County,* 19).

10. See, for example, the entries for May 3, 1862, and April 22, 1864, in George Lawrence Williamson's journal, in possession of Juanita Cody Williamson, Florence, S.C. Even in the coastal region, some planters who raised cotton and not rice as their commercial crop had small plots of rice for their own use. Thomas Chaplin, a European American cotton planter on Saint Helena Island, wrote in his journal on April 15, 1856: "Corn sprouting out. Planted rice through the lowest places in corn field" (Theodore Rosengarten, *Tombee: Portrait of a Cotton Planter, with the Journal of Thomas B. Chaplin (1822–1890)* [New York, 1986], 662).

11. Even when searching beyond the pine belt, I found few references to African Americans cultivating the small rice plots that were used for local consumption, and there was no suggestion that they may have contributed the knowledge on which the method was based. Years of searching produced fewer than a dozen references; three are included here. In 1854 De Bow wrote: "[Rice] is grown for domestic consumption in Mississippi, Florida, and Texas, and in nearly everywhere else where may be found a settlement of negroes who once lived in the rice region of the country [United States]" (De Bow, "Seacoast

Crops of the South," 592, 614). De Bow implies that African Americans who knew how to grow rice had learned the techniques on New World rice plantations. A 1905 United States government bulletin stated: "Four or five years ago the attention of the public [in Arkansas] was called to the small pieces of rice which were being grown in the low swampy places along the bayou. Among the most successful in this kind of rice culture were certain negroes, who raised enough rice for their own table" (Elwood Mead, ed., *U.S. Department of Agriculture Annual Report of Irrigation and Drainage Investigations, 1904* [Washington, D.C., 1905], 545). Milling, writing about the twentieth century, said that at Mechanicsville, a few miles north of Mars Bluff, African Americans "often raised a year's supply of rice on a small patch, threshing it out by hand in a primitive mortar" (Milling, "Mechanicsville," 150).

12. Phillips gives an example of the relative costs of corn and rice on a rice plantation. When broken rice was worth $2.50 per bushel, corn was worth $1.00 (Phillips, *American Negro Slavery,* 255).

13. Joseph A. Opala, conversation with author, Orangeburg, S.C., September 7, 1990. Opala was telling of the pride that people of Sierra Leone take in being rice growers.

14. That custom still prevailed in the author's home throughout the first half of the twentieth century. Interviews indicate that, during that fifty-year period, the eating habits of African Americans were shifting back toward rice. Archie Waiters had reported that former slaves Tom Brown and Alex Gregg preferred mush, grits, and cornbread to rice in 1920. On the other hand, African Americans who had been children in 1920—Matthew Williamson, Hester Waiters, Frances Johnson, and Rev. Frank Saunders—reported that they liked rice and ate it several times a week. None of them stated that they had to have it everyday as was the custom among European Americans at that time.

15. The practice of small-scale subsistence rice cultivation probably received more encouragement from pine belt landowners than from tidewater landowners, whose rice plantations produced all the rice they wanted for eating. For that reason, the custom of small-scale rice cultivation may have been more common in the pine belt.

16. A. Waiters interview, June 6, 1985, pp. 848–55. Uncle Sidney was Sidney Gregg, and Uncle Bubba was Spencer Gregg, both sons of Alex and Emma Gregg. Though Waiters was not able to walk to this rice field with me, he was able to define its location in relation to a ditch and a dryland field. I believe that I have the field's north/south location correctly identified; however, I am less confident about the location of the east/west boundaries, and I do not know the size of the field.

17. A. Waiters interview, July 31, 1987, pp. 1139–40. A record book from 1866 shows Gregg and Son purchasing a small amount of rice from a European American, so farmers at Mars Bluff may have been selling small amounts of

rice to Gregg for resale (J. Eli Gregg, Gregg and Son, Mars Bluff, Cotton and Corn Records, 1860–1867, bound MS vol., South Caroliniana Library, Columbia, S.C.).

18. It is not known exactly when the change was made to paying freedmen for rice. A letter written by J. Eli Gregg's son in 1867 indicates that Gregg still looked upon the rice being cultivated by freedmen as his own crop. Walter Gregg wrote to his wife, who was visiting in Columbia:

> Mars Bluff, South Carolina,
> October 11, 1867
>
> Dear Wife,
>
> While Father is off on my horse, I thought I would devote the time to writing you a short letter. Father has gone down in the swamp where the hands are cutting rice of which cereal he is making a very fine crop.
>
> (W. Gregg, letter to A. P. Gregg, October 11, 1867, photocopy in possession of Anna S. Sherman)

Also unknown is when African Americans started working on their own time to clear swampland for rice production and considering the fields they cleared as their own. The cotton and corn record book of J. Eli Gregg and Son for 1866 listed purchases of cotton from African Americans on the same page with a purchase of rice from a European American, so it seems that the transition to paying African Americans for rice could have been made easily as soon as the African Americans established that they were growing the rice on their own land (Gregg and Son, Cotton and Corn Records). Waiters said that Willie Scott was the last person he knew to grow rice and that he never sold any (A. Waiters interview, July 31, 1987, p. 1140). Based on changes in the local business situation and in national marketing patterns, I calculate that by the 1920s no merchant was buying rice raised at Mars Bluff.

19. Joe Opala told me that drivers on irrigated rice plantations exemplified the skill, status, and independence of rice growers everywhere. They had a body of knowledge that few European Americans ever mastered (Opala, conversation with author, September 7, 1990). James M. Clifton confirmed the prestige and autonomy of the skilled rice growers who became drivers in South Carolina (James M. Clifton, "The Rice Driver: His Role in Slave Management," *South Carolina Historical Society,* LXXXII [October, 1981], 331, 335, 338).

CHAPTER 12

1. A. Waiters interview, July 31, 1987, pp. 1117–18. Three times Waiters mentions throwing the leftover rice out to be eaten by chickens, dogs, or hogs.

No doubt in this case the rice was being thrown out as garbage for the animals to eat. However, it is worth observing that Herskovits wrote of throwing the first crops, especially of staples such as rice, out in the yard for the *Yorka* (the spirits of the ancestors) to eat. He quoted an African American in Surinam: "When you plant rice, the first rice you must not eat. Then you boil some. Then you throw (it) away on the ground." Herskovits explained: "This offering is given for the Earth Mother as well as for the ancestors, for both of these must eat of first fruits if the fields are to prosper" (Herskovits, *The New World Negro,* 316). See Appendix A on the expression "swallow your tongue."

2. Another interview with Waiters substantiates the claim that many of the old people did not eat rice; also, it elaborates on their practice of clearing land for rice:

> *Q.* When Frank Fleming was planting rice, he cleared all that land in there?
> *A.* He cleared up all—everybody what plant rice, they cleared their own land.
> *Q.* And Tom Brown, you reckon, cleared the back part?
> *A.* Every man what plant more rice clean up; so whoever didn't clean up some, he didn't plant more rice. That the way they done.
> *Q.* Now, your granddaddy had a lot of mortars and pestles. So you all must have eaten rice at his house?
> *A.* No, they didn't eat no rice. I tell you now when they make that rice. See they beat that rice. And they send that rice to Columbia, Charleston, and all that kind of mess there. They send that rice off.
> *Q.* They'd sell it to—
> *A.* Send it to these towns.
>
> (A. Waiters interview, July 31, 1987, pp. 1139–40)

3. Leland Ferguson, *Uncommon Ground: Archaeology and Early African America, 1650–1800* (Washington, D.C., 1992), 94. Recall the A. Waiters interview of June 5, 1985, quoted in Chapter 2. Charles Edwin Seagrave, in "The Southern Negro Agricultural Worker, 1850–1870" (Ph.D. dissertation, Stanford University, 1971), 56–57, cited three sources showing corn as the staple food of enslaved African Americans. One was Kenneth M. Stampp, who said, "A peck of corn meal and three or four pounds of salt pork or bacon comprised the basic weekly allowance of the great majority of adult slaves" (Stampp, *The Peculiar Institution* [New York, 1956], 282). The second was Clement Eaton, who said, "On Sunday the overseer apportioned the week's rations to the slaves, consisting of a peck of corn meal and three or four pounds of bacon to each" (Eaton, *The Growth of Southern Civilization,* 59). The third was Joe Gray Taylor, who said, "The usual practice throughout the ante-bellum period was to allow

each adult Negro a half a pound of pork and a quart of corn meal a day" (Taylor, *Negro Slavery in Louisiana* [Baton Rouge, 1963], 107. Even on rice plantations, corn played a role. Clifton wrote that at Gowrie, a rice plantation just above Savannah, weekly rations included "one peck of corn or corn meal. . . . Sometimes rice would replace the corn, but not often" (Clifton, *Life and Labor on Argyle Island,* xxxiii). Joyner gave numerous examples of rations in the Waccamaw tidewater rice region that consisted largely of rice. But "running close behind rice as a staple in All Saints Parish were grits and hominy, both derived from corn" (Joyner, *Down by the Riverside,* 96–99). An account of the relative costs of corn and rice on the Charles Manigault rice plantation shows why corn might have been issued as rations even on a rice plantation. A weekly peck of corn cost the plantation owner thirteen dollars per year. "In reward for good service, however, Manigault usually issued broken rice worth $2.50 per bushel, instead of corn worth $1" (Phillips, *American Negro Slavery,* 255). Even in the military, African Americans were restricted to corn. European Americans involved in military action against the Spanish in 1740 were given rice and corn, whereas African Americans were allowed only corn (Wood, *Black Majority,* 233). Charles Ball was part of a coffle that a slave trader took from Maryland to South Carolina about 1800. The slaves had corn to eat regularly along the way, but on arriving in South Carolina they were fed rice. Near Columbia, a local slave owner explained, "For some cause, which he could not comprehend, the price of rice had not been so good this year as usual, and that he had found it cheaper to purchase rice to feed his own *niggers* than to provide them with corn, which had to be bought from the upper country" ([Ball,] *Slavery in the United States,* 73, 80).

4. A. Waiters interview, December 28, 1977, p. 11. On James Island in 1883, rations were defined as "three pounds of bacon and one peck of grist a week" (State Board of Agriculture, *South Carolina Resources,* 30). Davis wrote about the pine belt after emancipation: "Negro plowmen on the agricultural farms and negro laborers on the turpentine farms received as part of their compensation a weekly ration. The ration of the farm worker was a peck of corn meal, three and a half pounds of salt bacon, a quart of molasses, and a pint of salt" (Davis, "A History of Florence," Chap. 16, p. 25).

5. I was encouraged to find that someone else has written about this trait. Karen Hess quoted from Alfred W. Crosby, Jr., *The Columbian Exchange* (1972), "It seems reasonable to say that human beings, in matters of diet, especially of the staples of diet, are very conservative, and will not change unless forced" (Karen Hess, *The Carolina Rice Kitchen: The African Connection* [Columbia, S.C., 1992], 56).

6. A. Waiters interview, July 31, 1988, pp. 1115–16.

7. *Ibid.,* June 5, 1985, p. 730.

CHAPTER 13

1. Frank Saunders, interview, August 17, 1987, p. 1266.

2. *Ibid.*, 1258, 1260–62, 1266, 1269. A portion of the interview shows one twentieth-century child's understanding of where the practice originated:

> And, well, we learned something about it, and I'm sure they made the biggest discovery because they the one plant.
> *Q.* . . . What. . . ?
> *A.* I said I enjoyed it. I enjoyed having the experience of knowing about the rice. But I say, I know they enjoyed it more because they were the one, as far as we were concerned, they discovered it. I don't know who they get it from. I don't know who or where they got the idea of it, but they certainly had it. And there was more than them. I wouldn't try to name, but there was more than them planting the rice. . . .
> *Q.* Tell me about clearing the land.
> *A.* That's what I'm telling you. . . . Right at the end of that cemetery across—that's what happened. They must have know something. Somebody must have told them how to do it. Because they did have that low land. That's one thing that Brown said. Brown said—told us, now, it could be on a low ground. That's exactly what it was. Some kind of muddy. (1261, 1269)

3. Paul Bohannan and Philip Curtin, *Africa and Africans* (Rev. ẹd.; Garden City, N.Y., 1971), 24–25; Hans Ruthenberg, *Farming Systems in the Tropics* (2nd ed.; Oxford, 1976), 22; Gordon Wrigley, *Tropical Agriculture: The Development of Production* (London, 1971), 28–29, 44–45, 82. These scholars explain that in the tropics the high temperature causes nitrogen and other essential nutrients to be lost from the soil. It also causes bacteria to work more rapidly on decayed plant remains, so there is little humus in the soil. The heavy rains leach the nutrients out of the topsoil and wash away exposed soil. The tremendous amount of sunlight causes all soil nutrients to be immediately recycled into plant growth.

Rodney, writing about Sierra Leone, gave a glimpse of African land-clearing activities.

> Although wet rice was important, the dry varieties were far more prevalent, and the most widespread form of agriculture practiced on the Upper Guinea Coast was the itinerant rotational one, which is still the most common. Essentially, this consists in winning from the forest

> each successive year enough ground on which to sow the necessary crops. . . . The period between March and June was one of great activity, when, by collective effort, certain areas selected by the tribal authorities were cleared of forest by cutting and burning. . . .
>
> The bulk of the heavy work of clearing the forest was undertaken by the men. The smaller trees were cut level with the ground, while the large tree stumps were left to be consumed by termites and fire. (Walter Rodney, *A History of the Upper Guinea Coast, 1545–1800* [London, 1970], 22–23)

4. Herskovits, *The New World Negro,* 58–59. The "weighting of the concerns of a people constitutes the focus of their culture. *Cultural focus* is thus seen to be that phenomenon which gives a culture its particular emphasis" (59).

5. *Ibid.,* 15. Continuing on the subject of our failure to appreciate the religious basis of African American life, Herskovits wrote:

> [We] fail to grasp that core of cultural reality in terms of the integrative norms which inhere in these practices, however vulnerable they may appear from the point of view of economic expediency. . . .
>
> Negroes have retained African religious beliefs and practices far more than they have retained economic patterns. But when we examine the patterns of African cultures, we find that there is no activity of everyday living but that it is validated by supernatural sanctions. And consequently, these figure far more in the total life of the people than does any other single facet of the culture such as those matters having to do with making a living, or family structure, or political institutions. (15, 58–59)

6. G. Howard Jones, *The Earth Goddess: A Study of Native Farming on the West Coast of Africa* (London, 1936), 157, 6–9, 48–49; Herskovits, *The New World Negro,* 15, 58–59, 316. Herskovits wrote of the close association between the earth and the ancestors, giving as an example a ritual in which one "offering is given for the Earth Mother as well as for the ancestors. . . . Nor is this the only instance where Earth and Ancestors are associated in ritual" (316).

7. Herskovits, *The New World Negro,* 84–85. Opala said that on a visit to South Carolina in 1988, the president of Sierra Leone took a glass of water and went outside and poured it on the earth while addressing his ancestors buried in that soil—soil he regarded as sacred. It is even possible that he was directly descended from someone buried in South Carolina soil, for many of the people who were brought to South Carolina as slaves had left children in Africa (Opala, conversation with author, September 7, 1990).

8. Paul Richards, conversation with author, Ames, Iowa, April 14, 1989.

CHAPTER 14

1. Litwack, *Been in the Storm So Long,* 21–27, describes the slaves' methods for learning about the progress of the war.

2. Joel Williamson, *After Slavery: The Negro in South Carolina During Reconstruction, 1861–1877* (Chapel Hill, 1965), 54. Du Bois quoted Congressman Thaddeus Stevens: "We have turned, or are about to turn, loose four million slaves without a hut to shelter them or a cent in their pockets. The infernal laws of slavery have prevented them from acquiring an education. . . . If we do not furnish them with homesteads and hedge them around with protective laws; we had better have left them in bondage" (Du Bois, *Black Reconstruction,* 265–66). Most people did not share Stevens' insight. According to Stampp, most people thought that emancipation was all that was necessary. "Even William Lloyd Garrison, the most militant of the old abolitionist leaders, was ready to dissolve the American Anti-Slavery Society after the Thirteenth Amendment had been adopted. To Garrison legal emancipation and civil rights legislation were the primary goals, and the economic plight of the Negroes concerned him a good deal less" (Kenneth M. Stampp, *The Era of Reconstruction, 1865–1877* [New York, 1966], 129). Willie Lee Rose quoted from Henry Ward Beecher's speech at the end of the war: "*One nation, under one government, without slavery.* . . . On this base reconstruction is easy, and needs neither architect nor engineer." Rose remarked, "The Port Royal veterans probably wondered where Beecher ever got an idea like that. If the Northern public should take Beecher's stand, the problems of social and economic reconstruction might never be met" (Willie Lee Rose, *Rehearsal for Reconstruction: The Port Royal Experiment* [Indianapolis, 1964], 344).

3. Wendell Phillips, quoted in James M. McPherson, *The Struggle for Equality: Abolitionists and the Negro in the Civil War and Reconstruction* (Princeton, 1964), 243.

4. Martin Abbott, *The Freedmen's Bureau in South Carolina, 1865–1872* (Chapel Hill, 1967), 52. Eric Foner wrote: "Access to land gave even the poorest blacks some measure of choice as to whether, when, and under what circumstances to present themselves in the labor market. But given the overall political economy of plantation regions, such autonomy tended to be at best defensive" (Eric Foner, *Nothing but Freedom: Emancipation and Its Legacy* [Baton Rouge, 1983], 36). King, writing about Reconstruction in the county where Tom Brown lived, said: "At the heart of any chance for blacks to successfully make the transition from slavery to freedom was the promise of '40 acres and a mule.' Political freedom without some economic freedom, as some in the government knew, was a cruel hoax" (King, *Rise Up So Early,* 60–61). Du Bois acknowledged that minimal basic capital was essential (Du Bois, *Black Reconstruction,* 222).

5. Williamson, *After Slavery*, 54. James M. McPherson demonstrated that "this hope—for 'forty acres and a mule'—was no delusion of ignorant minds" (James M. McPherson, *Ordeal by Fire: The Civil War and Reconstruction* [New York, 1982], 506–507). LaWanda Cox wrote: "Permanent confiscation of large southern plantations was implied both by Julian's southern homestead bill and by the act establishing the Freedmen's Bureau. This expectation, although never realized, was by no means fanciful in 1864" (LaWanda Cox, "The Promise of Land for the Freedmen," *Mississippi Valley Historical Review*, XLV [December, 1958], 432). Willie Lee Rose wrote: "Blacks' hope of a new beginning, with forty acres and a mule, hardly stemmed only from federal authorities. In fact, blacks had gleaned the idea from their former owners, during the war, that under the confiscation acts rebels' land would be seized and given to slaves. Planters had regarded this as the best reason to fight to the last ditch" (Willie Lee Rose, *Slavery and Freedom*, ed. William W. Freehling [New York, 1982], 82).

6. Du Bois, *Black Reconstruction*, 393; Ira Berlin *et al.*, eds., *The Wartime Genesis of Free Labor: The Lower South* (Cambridge, Eng., 1982), 106, Ser. I, Vol. III of Berlin *et al.*, eds., *Freedom: A Documentary History of Emancipation, 1861–1867*, 5 series projected.

7. Rose, *Rehearsal for Reconstruction*, 275, 282, 284–87.

8. *Ibid.*, 287–96; Claude F. Oubre, *Forty Acres and a Mule: The Freedmen's Bureau and Black Land Ownership* (Baton Rouge, 1978), 9–10.

9. Cox, "The Promise of Land," 429; Foner, *Reconstruction*, 70–71. Rose suggested another origin for the phrase "forty acres." The phrase "20 or 40 acres" was used in President Lincoln's December 31; 1863, provision for expansion of the lands available for preemption by African Americans (Rose, *Rehearsal for Reconstruction*, 285, 326–28).

10. Rose, *Rehearsal for Reconstruction*, 329–31. At one point African Americans were actually required to stake out land. Williamson wrote: "During the first month, settlement was voluntary. However, early in May, Negroes drawing government rations in Charleston were given ten days to find homesteads on the plantations or suffer loss of their allowances" (Williamson, *After Slavery*, 62).

11. Oubre, *Forty Acres and a Mule*, 20–21.

12. Williamson wrote: "Professor Martin L. Abbott, a careful student of the subject, estimated that Saxton eventually settled about 40,000 Negroes under the program" (Williamson, *After Slavery*, 63).

13. Oubre, *Forty Acres and a Mule*, 31, 49, 51.

14. Saxton, quoted *ibid.*, 49, 51. According to Abbott, Saxton's final recommendation was that Congress should buy the lands from their former owners, but no action was taken by Congress (Abbott, *The Freedmen's Bureau*, 59).

15. George R. Bentley, *A History of the Freedmen's Bureau* (New York,

1970), 89–102. In his autobiography General Howard described his reaction to the first presidential pardon:

> Surely the pardon of the President would not be interpreted to extend to the surrender of abandoned or confiscated property which in strict accordance with the law had been "set apart for refugees and freedmen." . . . Did not the law apply to all formerly held as slaves, who had become or would become free? This was the legal status and the humane conclusion. Then naturally I took such action as would protect the *bona fide* occupants, and expected the United States to indemnify by money or otherwise those Confederates who were pardoned; assuredly we would not succor them by displacing the new settlers who lawfully were holding the land.

Forced by President Johnson to order the restoration of the property to the former owners, Howard wrote: "My heart was sad enough when by constraint I sent out that circular letter; it was chagrined when not a month later I received . . . orders issued by President Johnson" stating that the lands under Sherman's Special Field Order No. 15 were also to be restored. Howard asked, "Why did I not resign? Because I even yet strongly hoped in some way to befriend the freed people" (Oliver Otis Howard, *Autobiography of Oliver Otis Howard, Major General United States Army* [New York, 1907], 234–35, 237–38).

16. Du Bois, *Black Reconstruction,* 602.

17. Quoted in Oubre, *Forty Acres and a Mule,* 53.

18. Rose, *Rehearsal for Reconstruction,* 355. McPherson wrote: "In February 1866, Congress passed a bill that extended the life of the Freedmen's Bureau and included a provision confirming the freedmen's possession for three years of lands occupied under the Sherman order. But Johnson vetoed the bill and Congress failed to pass it over the veto" (McPherson, *Ordeal by Fire,* 508). Hans L. Trefousse wrote that Carl Schurz toured the South and took reports to President Johnson saying that a redistribution of land was the best way to control vagrancy. President Johnson accepted Schurz's report reluctantly and refused to consider the course he suggested (Hans L. Trefousse, *Carl Schurz: A Biography* [Knoxville, 1982], 157–59).

19. Oubre, *Forty Acres and a Mule,* 22. Cox described the congressional negotiations that led to "the puzzle still unresolved: namely, that the Freedmen's Bureau Act provided for the *sale* of forty acre allotments and yet at the same time recognized that the United States might never have permanent title to the lands allotted, and specified that they were to be sold with 'such title as it could convey'" (Cox, "The Promise of Land," 432–33).

20. McPherson, *The Struggle for Equality,* 412. According to McPherson, a

bill that would have established a federal land commission died in committee. Willie Lee Rose explained why land was not made available for African Americans: "The real reason was that the number of persons who wanted an economic program to undergird emancipation was always very small" (Willie Lee Rose, "Jubilee & Beyond: What Was Freedom?" in *What Was Freedom's Price?* ed. David G. Sansing [Jackson, Miss., 1978], 14).

21. Du Bois, *Black Reconstruction,* 269–70. These congressional priorities were written into the Fourteenth Amendment to the Constitution (1868). One clause of that amendment involves the welfare of freedmen, whereas three clauses address the congressmen's concerns about who would control Congress.

22. John Richard Dennett, *The South as It Is, 1865–1866,* ed. Henry M. Christman (London, 1965), 187–89. Dennett wrote: "In Virginia and North Carolina I found but one Negro, an old man living near Charlotte, who entertained any expectation of this kind. In this State I have talked with the people of five plantations in Marion and Williamsburg districts, and all seemed to be fully persuaded that some such provision was to be made for them" (189). Bentley wrote that on New Year's Day, 1866, Main Street of Sumter, about fifty miles from Mars Bluff, was filled with freedmen expecting a gift of land (Bentley, *A History of the Freedmen's Bureau,* 82). According to Williamson, "The coming of the new year, 1866, saw many freedmen unwilling to enter into labor contracts." Williamson cited the January 13, 1866, letter of a Beaufort planter: "I have seen all the planters from Combahee, Ashepoo, Pon Pon, et cet, . . . not one has yet been able to make the negroes contract" (Williamson, *After Slavery,* 88).

23. McPherson, *Ordeal by Fire,* 510–11.

24. Edward Magdol, *A Right to the Land: Essays on the Freedmen's Community* (Westport, Conn. 1977), 188–89. General Howard wrote: "Again, I urged that to render any portion of the freedmen able to take advantage of the homestead law in Florida, Louisiana, Arkansas, or in other States where there were public lands, aid must be furnished the settlers in the way of transportation, temporary food, and shelter, and implements of husbandry. To render this relief offered effective, more time than our present law offered would be essential" (Howard, *Autobiography,* 243).

25. Francis Butler Simkins and Robert Hilliard Woody, *South Carolina During Reconstruction* (Chapel Hill, 1932), 24, 64, 72; Alrutheus Ambush Taylor, *The Negro in South Carolina During Reconstruction* (New York, 1924), 133–34. Taylor wrote that when the convention met, on January 14, 1868, the belief was so widespread that the United States government was going to give land to the freedmen that the convention was asked to help the Freedmen's Bureau persuade the freedmen that the federal government had no land for them and that they should sign labor contracts for 1868. Therefore, the constitutional convention passed a resolution:

> That this Convention do hereby declare to the people of South Carolina, and to the world, that they have no land or lands at their disposal, and in order to disabuse the minds of all persons whatever throughout the State who may be expecting a distribution of land by the Government of the United States through the Bureau of Refugees, Freedmen and Abandoned Lands, or in any other manner, that no act of confiscation has been passed by the Congress of the United States, and it is the belief of this Convention that there never will be, and that the only manner by which any land can be obtained by the landless will be to purchase it.
>
> (*Proceedings of the Constitutional Convention of South Carolina, 1868,* p. 213)

26. W. Gregg, letter to A. P. Gregg, December 26, 1867, photocopy in possession of Anna S. Sherman. (Both J. Eli Gregg and his son Walter owned the land.) That there was plenty of labor at Mars Bluff is significant, for a labor shortage was the chief weapon of the freedmen in the unequal contest between labor and landowners during Reconstruction. Foner wrote, "The freedmen utilized the labor shortage . . . to oppose efforts to put them back to work in conditions, especially gang labor, reminiscent of slavery" (Foner, *Nothing but Freedom,* 44). The records of Gregg and Son indicate that some Mars Bluff landowners may have allowed sharecropping in 1866—or maybe they were merely paying gang labor in cotton rather than in money. The September–November, 1866, books show that Gregg was buying cotton from both African Americans and European Americans: The names London, Tom, Jack, Richard, and Dana (assumed to refer to African Americans because no last names were given) are mixed in with the names Simon Deese, James Haywood, Tobias Bailey, Waterman Bailey, and Mrs. Bailey (assumed to refer to European Americans) (Gregg and Son, Cotton and Corn Records). These records indicate that at Mars Bluff the practice of not allowing slaves to have surnames was not immediately eliminated by emancipation. For more about the question of surnames, see Litwack, *Been in the Storm So Long,* 176, 247–49.

27. Richard Nelson Current, *Those Terrible Carpetbaggers* (New York, 1988), 222.

28. Carol K. Rothrock Bleser, *The Promised Land: The History of the South Carolina Land Commission, 1869–1890* (Columbia, S.C., 1969), 30–34, 47–65, 84. According to some accounts the land commissioner was paid $25,000 rather than $45,000. Bleser explained the discrepancy: the commissioner was offered $25,000 to resign and agreed to that amount. Then he insisted that he must also be paid $20,000 for his share of the Greenville and Columbia Railroad, so he was paid $45,000 altogether. When Francis L. Cardozo, an African American, assumed control of the land commission in 1872, it became more honest. By

then, however, most of the allocated funds had been spent, and there was little money for the work of the commission.

29. *Ibid.*, 77, 94. Hayne was later to be made land commissioner for the state.

30. *Ibid.*, 166–67. Bleser's figures show that the five counties with the largest acreages were Charleston, 25,501; Colleton, 12,894; Richland, 9,398; Chesterfield, 6,918; and Marion, 6,661. When I made a quick search of the Marion County deed books indexes for the period of Hayne's work in Marion County, I found only about 124 deeds recorded with the State of South Carolina as grantor. About 56 of these were specifically listed as from the Sinking Fund, the agency responsible for buying land and reselling it to African Americans; the other 68 entries were listed as from the State of South Carolina, so they may not have been the result of Hayne's work.

31. Deed from the Commissioners of Sinking Fund to Washington James *et al.* for 913 acres, March 4, 1884, Marion County, S.C., Deed Book MM, 274–75, in Records of the Budget and Control Board; Sinking Fund Commission, Public Land Division, Duplicate Titles, B, p. 424 and back of p. 424, South Carolina Department of Archives and History, Columbia, S.C. This was probably swampland.

32. Bleser, *The Promised Land,* 131, 141–43.

33. *Ibid.*, 158. In 1883 African American ownership of homes or land was 5 percent in Marion County and 2 percent in Darlington County (State Board of Agriculture, *South Carolina Resources,* 85). Reasons for unsuccessful purchase attempts included poor quality of land, over-pricing of land, a three-year residency requirement, the agricultural depression of the early 1870s, and after 1878, the requirement of full payment in two to four years (Bleser, *The Promised Land,* 65, 95, 130, 140–41, 157–58).

34. Du Bois, *Black Reconstruction,* 601, 603, 611. Simkins and Woody summed up the situation at the end of Reconstruction:

> It should not be assumed that the agricultural life of the state was entirely made over during Reconstruction. Some had predicted that the Negro would dominate the agricultural life of the state, becoming the owner of the soil. Others had predicted that the rising energy of the new white democracy would drive the black man from the state. But neither of these predictions materialized. Both races remained in about the same proportions and in about the same relative positions as before the war. The Negro continued to be the tiller of the soil and the white man continued to be the owner.
>
> (Simkins and Woody, *South Carolina During Reconstruction,* 265)

35. McPherson, *Ordeal by Fire,* 506. An 1866 report of the Joint Committee on Reconstruction states in part: "One universal opinion is that they [African

Americans] shall not be allowed to acquire or hold land. I have heard that expressed from the first. They say that unless Negroes work for them they shall not work at all" (Du Bois, *Black Reconstruction,* 369). An assistant commissioner for Mississippi and northeast Louisiana stated in July, 1865: "[Landowners] steadily refuse to sell or lease lands to black men. Colored mechanics of this city [Vicksburg] who have made several thousand dollars during the last two years, find it impossible to buy even enough land to put a house on, yet white men can purchase any amount of land" (Magdol, *A Right to the Land,* 139). A report to the House of Representatives stated: "When Thomasville, Georgia, freedmen wished to rent land in 1865, they were told that no black man would be able to have land by lease" (Magdol, *A Right to the Land,* 149). Simkins and Woody explained the European American position thus: "Landowners, because land was the principal resource which the war had left them, were forced to give greater attention to its exploitation" (Simkins and Woody, *South Carolina During Reconstruction,* 226). Williamson probably came close to speaking the European American mind on the subject of African Americans' ability to purchase land when he quoted Edmund Rhett's proposal for the South Carolina Black Code: "The general interest both of the white man and of the negroes requires that he should be kept as near to the condition of slavery as possible. . . . Negroes and 'their posterity' would be prohibited from acquiring '*Real Estate*'" (Williamson, *After Slavery,* 75).

At Mars Bluff, there were a few exceptions to the rule that there be no sales from European Americans to African Americans. An African American community of about fifteen houses known as Jamestown began in 1871 when Ervin James bought 109 acres at the edge of the Pee Dee Swamp for $700 (Deed from Eli McIsick and Mary E. Poston to Ervin James for 109 acres, January 23, 1871, Marion County, S.C., Deed Book DD, 494–95, in South Carolina Department of Archives and History, Columbia, S.C.). In 1891, McIsick's daughter and son-in-law sold an additional 109 acres to six members of the James family for $500 (Deed from Sarah and J. A. Grice to Sidney James *et al.* for 109 acres, March 26, 1891, Florence County, S.C., Deed Book E, 413–15, in South Carolina Department of Archives and History, Columbia, S.C.). I was told by an older European American who had lived at Mars Bluff as a child that there was community disapproval of this sale and that therefore the family who made the sale was obliged to leave Mars Bluff; they went to Georgia. This was merely hearsay; however, even if the report is not true, it reflects the prevailing attitude about such sales (Catherine Edwards, conversation with author, Florence, S.C., early 1970s). Another report came from Rubin Peterson: "When I was a child in Jamestown, Jack James told me that Jamestown people had gotten that land because they worked for a man for a long time. He felt he owed each family a house and two acres of land. They said he deeded it to them, but I don't believe anybody had their own deed" (Rubin Peterson, interview, February 8, 1986, pp. 1004–1005). Other sales to African Americans at Mars Bluff included three to Anthony H. Howard: thirty-five acres in 1867, ninety-nine acres in 1871, and twenty-seven acres in 1875 (Marion County, S.C., Deed Books EE, 347, FF, 167, and GG, 674, at Marion County Court House, Marion, S.C.).

36. Sidney Andrews, *The South Since the War* (New York, 1969), 206. Andrews also reported: "Official reports from Marion, Darlington, and Williamsburg Districts represent the negroes as quiet, well-disposed, and generally at work for mere starvation wages. My own observations in these three districts pretty fairly confirm these reports, and furnish some clew to the bearing of the whites" (203).

37. Frederick Douglass described the condition of the African American freedman: "He had none of the conditions for self-preservation or self-protection. He was free from the individual master, but the slave of society. He had neither money, property, nor friends. He was free from the old plantation, but he had nothing but the dusty road under his feet. He was free from the old quarter that once gave him shelter, but a slave to the rains of summer and to the frosts of winter. He was, in a word, literally turned loose, naked, hungry, and destitute, to the open sky" (Frederick Douglass, *Life and Times of Frederick Douglass, Written by Himself: His Early Life as a Slave, His Escape from Bondage, and His Complete History* [Rev. ed., 1892; rpr. New York, 1962], 377).

38. A. Waiters interview, July 31, 1987, p. 1134.

39. Williamson wrote that this sort of control originated in 1865. He told about Sumter, a village about fifty miles from Mars Bluff: "Occasionally, landowners and employers entered into 'gentlemen's agreements' to refuse to rent land to Negroes and to minimize other concessions generally. On December 23, 1865, only two days after General Saxton had met the freedmen and planters from three districts in the village of Sumter to discuss contract arrangements for the coming year, the planters held a private meeting in the same place and agreed 'not to hire their neighbor's negroes, or rent any land to them.' More often, agreement was achieved informally within the tightly knit employer class" (Williamson, *After Slavery,* 99–100). Such restrictions apparently did not exist in earlier years. In 1822 and in the 1830s, William Ellison, "a free man of color" who built and repaired cotton gins, was able to buy prime land in Sumter (Michael P. Johnson and James L. Rourk, *Black Masters: A Free Family of Color in the Old South* [New York, 1984], 89–93).

40. Du Bois, *Black Reconstruction,* 367–68.

41. A. Waiters interview, December, 1977, p. 17. This was the first time I had heard the figure ten acres used rather than the common "forty acres and a mule."

42. A. Waiters interview, June 5, 1985, pp. 728–29.

43. Foner, *Reconstruction,* 106.

44. In an interview quoted earlier, Archie Waiters spoke as if European Americans might have shared that view. Talking about Alex Gregg's sharecrop-

ping, probably in the twentieth century, Waiters said: "He had a three-horse farm. He had twenty acres of cotton or something like that—corn, rice. Course, rice don't count because they give you that. Just corn and cotton. Most of it cotton" (A. Waiters interview, June 5, 1985, p. 730). That the landowner did not charge a share of the crop as rent on the rice land suggests that the landowner did not consider the rice field his, though he knew he owned the entire area. Perhaps European Americans had adopted the African concept of land ownership in regard to the swampland rice fields: the person who cleared the land owned it for as long as he wanted to use it.

45. Jones, *The Earth Goddess,* 24.

46. Melissa Leach, letter to author, May 23, 1989.

47. Richards, conversation with author, April 14, 1989. Franklin wrote about the African view of land ownership: "The land was considered so important to the entire community that it belonged not to individuals but to the collective community. . . . Individuals or groups of persons could obtain the right to use a given parcel of land, but such permission did not carry with it the right of alienation or any other form of disposition. When use was not made of the land it reverted to the collective domain" (Franklin, *From Slavery to Freedom,* 25–26). Bohannan and Curtin wrote of the difference in the European American and the African views of land: "Land . . . is a 'thing' that modern Westerners cut into pieces that they call parcels which they can then buy and sell on the market." This is in contrast to the African, who has "farm tenure" but not "land tenure"—he has the right to cultivate a piece of land, but he has no right of ownership of the land itself (Bohannan and Curtin, *Africa and Africans,* 120, 124).

48. Franklin wrote, "African survivals in America also suggest a pronounced resiliency of African institutions" (Franklin, *From Slavery to Freedom,* 40).

49. Brown's land clearing, which was based on African traditions and influenced by his experiences in the New World, is best appreciated in the light of Herskovits' explanation of the evolution in the search for Africanisms. At the beginning of the twentieth century, there was no search for Africanisms in the New World. The consensus was that Africa had no culture of its own and consequently was incapable of making any contribution to the culture of the New World. Then occasional historians and anthropologists began suggesting that there were Africanisms in the New World, and they tried to find pure African practices to prove this theory. Herskovits saw that this was too narrow an approach. He wrote:

> *It came to be recognized that the problem was vastly more complex than a statement drawn in terms merely of the presence or absence of Africanisms in the New World. The problem . . . was the manner in which elements of European, African, and . . . American Indian cultures had exerted mutual*

influences. . . . Concepts of retention and reinterpretation reorient the entire field, raising research from the elementary level of description and comparison to that of the analysis of process. . . . They replace the initial questions, "What?" and "Where?" with the more penetrating "Why?" and "How?"
(Herskovits, *The New World Negro,* 23, 36–37, 57)

Epilogue

1. Pierre Poivre, *Travels of a Philosopher; or, Observations on the Manners and Arts of Various Nations in Africa and Asia* (Baltimore, 1818), 8–9.

2. Leon Festinger wrote: "Given that a cognition is responsive to 'reality' . . . if the behavior of the organism changes, the cognitive element or elements corresponding to this behavior will likewise change" (Leon Festinger, *A Theory of Cognitive Dissonance* [Stanford, 1957], 19). David G. Myers wrote: "Festinger's *cognitive dissonance* theory . . . assumes we feel tension ('dissonance') when two of our thoughts or beliefs ('cognitions') are psychologically inconsistent—when we recognize that they don't fit together. . . . We adjust our thinking to reduce the tension" (David G. Myers, *Social Psychology* [New York, 1983], 58).

3. Theodore Rosengarten made a similar discovery when he set out to write a book based on interviews. He wrote: "I returned in June with a hundred pages of questions to ask Shaw. It became clear during our first session that I'd never get to a fraction of them. It would have taken years; moreover, my prepared questions distracted Shaw from his course. . . . I learned how to listen" (Theodore Rosengarten, *The Life of Nate Shaw* [New York, 1974], xviii).

Appendix A

1. Melville J. Herskovits, "On the Provenience of New World Negroes," *Social Forces,* XII (December, 1933), 247, 252, 260. Herskovits wrote that the investigation of historical and ethnological data when combined could yield information about the provenance of African Americans. He saw the historical evidence as pointing to limited areas of West Africa and the Loango and Angola area.

2. Joko Sengova, conversation with author, Charleston, S.C., September 18, 1990.

3. C. Waiters interview, March 9, 1990, n.p. Herskovits mentions "the passing of small children over the coffin" in a discussion of West Africa, implying that this was a West African custom (Herskovits, "On the Provenience of New World Negroes," 261).

4. Nichols, ed., *The Last Miles of the Way*, 17.

5. John Gregg, Appraisement of Estate.

6. Winifred Kellersberger Vass, *The Bantu Speaking Heritage of the United States* (Los Angeles, 1979), 106. Vass's list of sources for words is headed "Vocabulary List of Possible Bantu Origin"; consequently, in every case in which I cite Vass, she has only suggested the possibility of Bantu origin, not affirmed it. Dalgish wrote that the word *biddy* is related to the Chiluba word *bidibidi*, meaning "small yellow bird" (Gerard M. Dalgish, *A Dictionary of Africanisms: Contributions of Sub-Saharan Africa to the English Language* [Westport, Conn., 1982], 24). Turner wrote that the word is Kongo and means "a bird" (Turner, *Africanisms*, 191).

7. *Webster's Third New International Dictionary* (unabridged), s.v. *biddy*.

8. Vass, *The Bantu Speaking Heritage*, 106. *Webster's Third* gives no derivation and defines *boogerman* as "a monstrous imaginary figure used in threatening children" (s.v. *boogerman*).

9. C. Waiters interview, March 9, 1990, n.p.

10. Robert Farris Thompson, *Flash of the Spirit: African and Afro-American Art and Philosophy* (New York, 1983), 142–44.

11. L. Coker, interview, March 14, 1984, p. 183; Turner, *Africanisms*, 197. Turner wrote of the similarities between the Gullah word "*kuta* 'tortoise'" and the "Bambara and Malinke [French West Africa], *kuta* 'tortoise'; Efik [southern Nigeria], *ikut* 'tortoise'; Djerma [French West Africa], *ankura* 'tortoise'; Tshiluba [Belgian Congo], *nkuda* 'tortoise, turtle.'" *Webster's Third* acknowledges the African origin of the word *cooter* (s.v. *cooter*).

12. Vass, *The Bantu Speaking Heritage*, 108.

13. A. Waiters interview, December 29, 1977, p. 26.

14. Judith Wragg Chase, *Afro-American Art and Craft* (New York, 1971), 55.

15. A. Waiters interview, July 31, 1987, p. 1116; Vass, *The Bantu Speaking Heritage*, 108. *Geechy*, as used by Waiters to mean "ticklish," is not to be confused with the proper noun *Geechee*, which refers to a dialect similar to Gullah. *Webster's Third* does not list *geech* or *geechy* and gives a New World derivation for *Geechee*, saying that it is "1. A dialect containing English words and words of native African origin spoken chiefly by the descendants of Negro slaves settled on the Ogeechee River in Georgia—compare Gullah. 2. A Geechee-speaking Negro" (s.v. *Geechee*). Vass, in discussing the origin of the word *Gullah*, wrote that "*Geechee*, another coastal dialect term, is thought to be derived from *Kisi*, a western Atlantic language from Guinea" (Vass, *The Bantu Speaking Heritage*, 32).

16. Saunders interview, August 17, 1987, p. 1270; Thompson, *Flash of the Spirit*, 105. Hurston wrote: "It will be noted how frequently graveyard dust is required in the practice of hoodoo, goofer dust as it is often called. . . . Dirt from sinners' graves is supposed to be very powerful, but some hoodoo doctors

will use only that from the graves of infants" (Hurston, "Hoodoo in America," 397).

17. C. Waiters interview, March 9, 1990, n.p.; Thompson, *Flash of the Spirit,* 134.

18. After telling of the contributions West Africa made in the development of agriculture, Murdock wrote, "In the realm of domesticated animals Negro Africa has not made a comparable contribution, the guinea fowl (*Numida meleagris*) constituting its sole original domesticate" (Murdock, *Africa: Its People,* 70).

19. Zanders interview, August 4, 1987, pp. 1202–1203.

20. Turner, *Africanisms,* 194. Turner wrote that the word is similar to words used by the Tshiluba in the Belgian Congo and the Umbundu in Angola. *Webster's Third* gives the origin as "[American French (Louisiana) *gombo,* of Bantu origin: akin to Umbundu *ochinggômbo* okra, Tshiluba *chinggômbô*]" and defines it as "the okra plant or its edible pods" or "a soup thickened with okra pods" (s.v. *gumbo*).

21. Margaret Washington Creel, *"A Peculiar People": Slave Religion and Community-Culture Among the Gullahs* (New York, 1988), 6.

22. *Webster's Third:* "hoppin john, also hopping john, usually capital J: a stew made with cowpeas, rice, and bacon or salt pork esp. popular in the southern states and traditionally served on New Year's Day" (s.v. *hoppin john*). I do not believe that the author of those lines had ever seen hoppin John; it bears no resemblance to a stew.

23. Murdock, *Africa: Its People,* 68.

24. F. R. Irvine, *A Text-Book of West African Agriculture: Soils and Crops* (London, 1935), 35–36. Irvine described both mixed cropping and intercropping in Africa. Dorothy Davis described intercropping at Mars Bluff, a practice that continued well into the first half of the twentieth century. The corn grew alone until it was lay-by size. When the corn was being plowed for the last time, peas were planted. The pea vines climbed the corn stalks, and the peas were picked well before the corn was harvested (Dorothy Hicks Davis, telephone conversation with author, December, 1988). I had always assumed that this was a universal practice, but Gerald McGrane told me that it was not done in Iowa (Gerald McGrane, telephone conversation with author, December 15, 1988). Coon and Ulrich Bonnell Phillips gave evidence that the practice was widespread in the South. Coon referred to early sources when he described English settlers in South Carolina planting "bonavis" peas in corn fields (Coon, "The Development of Market Agriculture," 133). Phillips wrote: "Peas . . . often called cowpeas . . . are planted in the cornfield when the corn is 'laid-by'" (Ulrich Bonnell Phillips, *Life and Labor in the Old South* [Boston, 1929], 4). European American farmers probably adopted the practice of intercropping corn and peas because they needed large quantities of peas, a mainstay in the diet of African Americans and a good source of protein. Henry Brown of

Charleston recalled: "My grandfather and grandmother were grown when they came from Africa, and were man and wife in Africa. . . . When the boats first came in from Africa with the slaves, a big pot of peas was cooked and the people ate it with their hands right from the pot" (Henry Brown, interview by Jessie A. Butler, n.d., in Rawick, ed., *The American Slave,* Vol. II, Pt. 1, pp. 118, 119–20).

25. Herskovits, *Myth of the Negro Past,* 149. Writing about the use of kerchiefs in Paramaribo, Herskovits said that the various ways that kerchiefs were tied sent messages. Some of the messages were "Wait for me at the corner," "I will not return," "You love the man, but he does not love you (as yet)," and "Virgin." Those carefully arranged kerchiefs were for street wear. "At home, a woman wears no kerchief at all, or she may wear an old unstarched bit of cloth" (Melville J. Herskovits and Frances S. Herskovits, *Surinam Folk-Lore* [New York, 1936], 7–9). This latter practice corresponds to the practice at Mars Bluff.

26. A. Waiters interview, June 6, 1985, p. 833.

27. Turner, *Africanisms,* 198.

28. Williamson interview, May 14, 1985, p. 687. Williamson's use of two verbs ("My mother told me, say . . .") is characteristic of the speech of many African Americans at Mars Bluff. Could it be related to what Molefi Asante refers to as the "dissecting quality of African linguistic expression," where every detail is expressed by a special verb? See Molefi Kete Asante, "African Elements in African-American English," in *Africanisms in American Culture,* ed. Joseph E. Holloway (Bloomington, Ind., 1990), 27.

29. Sengova, conversation with author, September 18, 1990; Alpha Bah, conversation with author, Charleston, S.C., September 18, 1990.

30. Vass, *The Bantu Speaking Heritage,* 112. The word *maum* is not found in *Webster's Third.*

31. Turner, *Africanisms,* 117.

32. *Webster's Third,* s.v. *buckra. Webster's* says the word is derived from Ibibio and Efik. John Bennett, writing about Gullah, said, "Among the few strange words of dubious [the word *dubious* has an X over it] African extraction, the most familiar [is] *buckra,* sometimes used as the Virginia negro uses *quality*" (John Bennett, undated file of notes [*ca.* 1900–30], File 21-111, Folder 20, South Carolina State Historical Society, Charleston, S.C.).

33. Turner, *Africanisms,* 72, 174.

34. Ida Zanders, letter to author, December 6, 1987; John W. Barber, comp., *A History of the Amistad Captives* (1840; rpr. New York, 1969), 13; Paul Richards, letter to author, [March ?], 1989.

35. Sengova suggested that *Tepee* might be a Kru name (Sengova, conversation with author, September 18, 1990).

36. Most of these names were taken from the author's notebooks of interviews. Some were given in interviews by persons previously identified; others

who supplied names were Annie Coker, Dorothy Smalls Williams, and Rubin Peterson. Three other sources for names were as follows: the name *Epshen* from Williamson family data, obtained from Matthew Williamson and filed in author's roots notebook on Williamson family-tree page; the name *Cuar* from James Gregg, Appraisement of Estate; the name *Tepee* from John Gregg, Appraisement of Estate.

37. C. Waiters interview, September 21, 1990, n.p.; Turner, *Africanisms,* 199; Sengova, conversation with author, September 18, 1990.

38. A. Waiters interview, February 26, 1986, p. 934.

39. C. Waiters interview, June 29, 1988, pp. 1357–58.

40. Guy B. Johnson, *Folk Culture on St. Helena Island, South Carolina* (Chapel Hill, 1930), 57.

41. Hazel Carter, telephone conversation with author, December 2, 1990, letter to author, December 2, 1990. Carter said that the Kongo word from which *pinda* is derived is *mphiinda,* commonly spelled *mpinda*. The Kongo and Mbundu word from which *goober* is derived is *nguba*. Carter explained that the prefix *ki,* as used by Turner in the word *Kimbundu,* refers to the language; the name of the people is Ambundu. Turner wrote of the similarities between the "Gullah *pinda* 'peanut'" and the "Kongo *mpinda* 'peanut'"; also between the "Gullah *guba* 'peanut'" and the "Kimbundu (Angola), *nguba* 'peanut'"; and between the "Umbundu (Angola), *olungupa* 'peanut'" and the "Kongo (Angola) *nguba* 'kidney'" (Turner, *Africanisms,* 194, 199).

42. Opala stressed that the tracing of the word *pinda* can be used to establish a probable connection of some kind, perhaps very indirect, between Mars Bluff people and African people who used a similar word. It does not rule out the possibility, however, that there are also Mars Bluff connections with African people who used words similar to *guba* (Opala, conversation with author, September 7, 1990).

43. A. Waiters interview, June 5, 1984, p. 719.

44. St. Clair Drake, *The Redemption of Africa and Black Religion* (Chicago, 1970), 13. Phillips wrote about the middle zone of the Slave Coast: "Hence were brought Eboes of jaundice tinge in eyes and skin, with prognathous faces and mournful natures inclining them in duress to seek death by their special device of 'swallowing their tongues'" (Phillips, *Life and Labor,* 190).

45. Sengova, conversation with author, September 18, 1990.

46. According to Opala, in Sierra Leone a palm-leaf brush broom is used. Because the broom is short, using it requires leaning over. The people of Sierra Leone lean from the waist when sweeping just as they do when hoeing with their short-handled hoes. When asked why they didn't put long handles on brooms and hoes so they would not have to lean over, Opala said that these people have plenty of material to construct handles if they want them, but that

just as each people has its own oral language, each has its own body language, and leaning from the waist is African body language (Opala, conversation with author, September 7, 1990).

47. Jane Vernon, who saw yards being swept with brush brooms in Nigeria, suggested that because a typical residence is a compound or cluster of small one-room buildings sharing a common yard where many household tasks are performed, the whole compound is considered part of the living space. Given this arrangement, keeping a clean yard is probably synonymous with keeping a clean house (Jane Vernon, telephone conversation with author, August 21, 1988, letters to author, August 24, September 19, 1988).

48. Sengova, conversation with author, September 18, 1990.

49. Opala, conversation with author, September 7, 1990.

50. Bah, conversation with author, September 18, 1990.

51. Williamson interview, May 14, 1985, pp. 660, 662, 667.

52. Saunders interview, August 17, 1987, p. 1262.

53. Lydia Parrish, *Slave Songs of the Georgia Sea Islands* (Hatboro, Pa., 1942), 232.

54. I have no examples of these words in my notebooks of interviews because in transcribing the interviews I followed the rule that all words should be written using standard English spelling, not the dialect spelling that has been traditionally used.

55. Clifford N. Fyle and Eldred D. Jones, *A Krio-English Dictionary* (Oxford, 1980), 380. Vass gave many spellings for this word: *hunnah, hoonah, oonah, une, unna, wunna* (Vass, *The Bantu Speaking Heritage,* 110, 112, 115). Turner said that the word *unu* is found in Ibo in southern Nigeria (Turner, *Africanisms,* 203). It is quite possible that the word *una* was used by Mars Bluff African Americans and that my ear never distinguished it from *yunna.*

APPENDIX B

1. Putting a sheaf of rice in a burlap bag for flailing seemed to me to be a strange method, so I was interested to learn from Opala that he had been told by an elderly African American in Riceboro, Georgia, that he had used the burlap bag method for threshing (Opala, conversation with author, September 7, 1990).

2. Richards, conversation with author; April 14, 1989; Paul Richards, *Coping with Hunger: Hazard and Experiment in an African Rice-Farming System* (London, 1986), 144–45.

3. Heyward, *Seed from Madagascar,* 21. Dennis T. Lawson gave a more detailed description: "Winnowing houses were about ten feet square and situated

on four columns about fifteen feet from the ground. A stairway was attached and a grating was placed in the middle of the floor of the house. When the wind was blowing strongly, the threshed seed was dropped through the grating to allow the chaff to be separated from the seed. . . . When the seed landed on the packed clay floor beneath the winnowing house laborers swept away any remaining waste or dirt—tailing, from the grain" (Dennis T. Lawson, *No Heir to Take Its Place: The Story of Rice in Georgetown County, South Carolina* [Georgetown, S.C., 1972], 8–9. I am indebted to Peter Wood for calling to my attention an early twentieth-century painting by Alice Ravenel Huger Smith entitled *The Threshing Floor with a Winnowing-House*. It shows that "winnowing houses" were not necessarily houses. Smith's is simply a platform on a tall, exposed scaffolding. See Alice Ravenel Huger Smith, *A Carolina Rice Plantation in the Fifties: Thirty Paintings in Water-Colour* (New York, 1936), n.p.

4. Johnson interview, March 12, 1984, pp. 185D–185F; Agricultural Extension and Research Liaison Services, Ahmadu Bello University, *Recommended Practices for Swamp Rice Production* (Zaria, Nigeria, 1981). In discussing the threshing methods used on a rice plantation, Sass recommended whipping the sheaf against something. He said that even after threshing mills were available, "the flail . . . continued to be employed for threshing of rice to be used for seed, because grain threshed in a mill seldom gave a good stand. An even better method of obtaining good seed rice was by the whipping off of the grain against a barrel, plank or log" (Sass, "The Rice Coast," 22).

5. A. Waiters interview, February 20, 1982, pp. 91–95. A horizontal pole three feet above the ground, resembling a carpenter's horse, was also described by Maggie Black, a contemporary of Alex Gregg in Marion County:

> Honey, peoples hadder work dey hand fa eve'yt'ing day hab mos' den. Dey grow dey own rice right dere on de plantation in dem days. Hadder plant it on some uv de land wha' wuz weter den udder land wus. Dey hadder le' de rice ge' good en ripe en den dey'ud cut it en hab one uv dem big rice whipping days. Heap uv people come from plantation aw 'bout en help whip dat rice. Dey jes take de rice en beat it 'cross some hoss dat dey hab fix up somewhey dere on de plantation. Honey, dey hab hose jes lak dese hoss yuh see carpenter use 'bout heah dese days. Dey'ud hab hundreds uv bushels uv dat rice dere. Den when dey ge' t'rough, dey hab big supper dere fa aw dem wha' whip rice. Gi'e em aw de rice en hog head dey is e'er wan'. Man, dey'ud hab de nicest kind uv music dere. Knock dem bones togedder en slap en pat dey hands to aw kind uv pretty tune.
>
> (Maggie Black, interview by A. Davis, June 21, 1937, in Rawick, ed., *The American Slave,* Vol. II, Pt. 1, pp. 58–59)

6. S. A. Knapp, "Rice," *Cyclopedia of American Agriculture* (New York, 1907), II, 535. Chan Lee described how subsistence farmers in Louisiana threshed rice. Some of these practices may have had an African origin.

> Several methods of varying complexity were known. The most simple method . . . was to beat the cut and dried rice with ordinary sticks. Some grasped the rice bundle and beat it on a log, a barrel, or some similar object, until most of the grain was removed from the straw. The bundles were then beaten with sticks to remove the remaining grains. Finally, the rice was winnowed with Indian basketry trays made of cane when there was breeze sufficient to blow the chaff away. Besides these simple techniques were animal treading and the use of true flails.
>
> (Chan Lee, "A Culture History of Rice with Special Reference to Louisiana" [Ph.D. dissertation, Louisiana State University, 1960], 109–10)

7. Irvine, *A Text-Book of West African Agriculture,* 95–96. Richards wrote that at present the Mende of Sierra Leone "tread rather than flail their rice" (Richards, letter to author, [March?], 1989).

8. Littlefield has suggested that one way to establish connections between Africa and African Americans would be by tracing a concrete technical procedure to a specific part of Africa (Littlefield, *Rice and Slaves,* 175–76).

9. Charles Gayarré, *History of Louisiana* (1885; rpr. New York, 1972), I, 290.

10. Lee, "A Culture History of Rice," 110–11. Lee includes a photograph of a wooden mortar and pestle from the lower part of Bayou du Large, Terrebonne Parish. The shape of the mortar is difficult to judge, though it appears to have a slight indentation midway between top and bottom. The pestle is markedly different from one Archie Waiters made to illustrate how pestles were made at Mars Bluff. Although it is difficult to say conclusively from the photograph, the pestle seems to be completely lacking a bulbous tip. It is of uniform diameter (about three or more inches) except at the center, where it seems to be indented as if for a hand grip.

The providence-rice area referred to in the quote is southwestern Louisiana. See Appendix E. *Encyclopedia Americana* defines *providence rice* as rice "grown in small patches, entirely with hand labor, without irrigation, and it was referred to as providence rice because farmers used rainwater collected in low-lying areas for their plantings" (*Encyclopedia Americana,* international ed. [1985], s.v. *rice*). Throughout most of the United States outside of Louisiana, such rice is called *upland rice.* Davis pointed out what a misnomer that was: "While called upland rice, it is really grown in the low ground spots such as drained ponds and sections of swamp found in the upland areas" (Davis, "A History of Florence," Chap. 15, p. 17).

11. Wood, *Black Majority,* 61–62; Joyner, *Down by the Riverside,* 58.

12. Littlefield, *Rice and Slaves,* 105–106. From Robert C. West's statement about a "Negroid area" of Colombia, South America, it can be seen that identifying the Native American and African American influences in South Carolina is relatively easy compared with the task of sorting out the history of the use of the wooden mortar and pestle by Native Americans and African Americans in Colombia. West wrote: "For local consumption rice is hand-hulled in a wooden mortar and pestle (pilón), the instrument used in the highlands to decorticate and pulverize maize. It is not clear whether the mortar and pestle were employed in the lowlands before the beginning of rice cultivation [1858]. Today Indians and Negroes never prepare maize in the wooden mortar but use the mealing stone and muller" (Robert C. West, *The Pacific Lowlands of Colombia: A Negroid Area of the American Tropics* [Baton Rouge, 1957], 149).

13. Joyner, *Down by the Riverside,* 58; Chase, *Afro-American Art and Craft,* 59–60; Dale Rosengarten, "Lowcountry Baskets: A Place in the Sun," *Carologue* (January–February, 1987), 8.

14. Chase, *Afro-American Art and Craft,* 60. Parrish's pictures show the difference between the shapes of mortars in Africa and Georgia (Parrish, *Slave Songs of the Georgia Sea Islands,* facing pp. 46, 47).

15. Opala, conversation with author, September 7, 1990.

16. Iliya Likita, conversation with Jane Vernon, Bambur, Nigeria, August, 1988.

17. Parrish, *Slave Songs of the Georgia Sea Islands,* 227–32.

18. Johnson interview, March 12, 1984, pp. 185h–185i.

19. H. Waiters interview, August 14, 1987, p. 1221. Opala said that an elderly African American near Darian, Georgia, who practiced subsistence rice cultivation until the 1950s told of using corn shucks for polishing rice (Opala, conversation with author, September 7, 1990).

20. William Henry Davis, interview by A. Davis, August 20, 1937, in Rawick, ed., *The American Slave,* Vol. II, Pt. 1, pp. 309–10.

21. Dale Rosengarten, *Row upon Row: Seagrass Baskets of the South Carolina Lowcountry* (Columbia, S.C., 1986), 5, 25, 35.

22. Lee, "A Culture History of Rice," 109, 112. Lee includes a photograph of two tray-type baskets with the caption, "Winnowing baskets used by Indian swamp-rice farmers near Dulac, Terrebonne Parish." Both baskets were right-angle weaving; one was made of lightweight material and resembled a basket I obtained from Zaire about 1960; the other was of heavier material and looked more like split-oak construction. From all appearances the construction of these

baskets from Louisiana and Zaire contrasts sharply with that of the coiled-grass baskets of the South Carolina low country, as pictured in Rosengarten (Rosengarten, *Row upon Row,* 5, 8–9, 15, 20, 24).

Appendix C

1. Jones, "Upland Rice," 1.

2. Brothers Jim and Willie Hayes had their fields beside Polk Swamp Canal and beside a major ditch that drained into the canal. Alex Gregg's field, on a perennial pond site, was close to the canal but not close enough to draw water from it. The field was surrounded on two sides by high land with a water table of four to six feet and some land that had a water table that ranged from six feet to too deep for field mapping (*Soil Survey of Florence and Sumter Counties, South Carolina* [Washington, D.C., 1974], 80–84, soil maps 6, 11).

3. Richards, *Coping with Hunger,* 58. In conversation, Richards pointed out how different conditions in South Carolina were from those in rain-fed rice-growing areas of West Africa, especially in the amount of rainfall. The fact that African Americans would even try to grow rice at Mars Bluff was a tribute to their adaptiveness.

4. Pantego soils are "nearly level and very poorly drained. . . . The surface layer is black loam about ten inches thick. The next layer is friable sandy clay loam that extends to a depth of about sixty inches. . . . Permeability is moderate, runoff is very slow or the soils are ponded." In this section of Mars Bluff, Pantego is the soil that has the highest water table. It is found along water courses, sometimes as much as a half-mile wide, but usually much less. Barth soils are "nearly level and moderately well drained. . . . The surface layer is very dark grayish-brown loamy sand about eight inches thick. Below this is a layer of very friable loamy sand about twenty-six inches thick. . . . Permeability is moderately rapid to rapid, but an occasional high water table impedes permeability" (*Soil Survey of Florence,* 11, 34).

5. A. Waiters interview, June 28, 1988, p. 1320.

6. H. Waiters interview, August 14, 1987, p. 1220; Johnson interview, March 12, 1984, p. 185b; Smith, telephone conversation with author, March 13, 1984.

7. Some dryland rice growers used Coxville or Goldsboro soils. Coxville soils "are nearly level and poorly drained. . . . The surface layer is very dark gray fine sandy loam about six inches thick. The next layer is dominantly gray and about fifty-nine inches thick. The upper sixteen inches is sandy clay loam that is mostly firm; the lower forty-three inches is very firm sandy clay grading to firm as depth increases. . . . Permeability is moderately slow, and runoff is

ponded or slow." Goldsboro soils "are deep, moderately well drained, and nearly level. . . . The surface layer is dark-gray loamy sand about seven inches thick. The subsurface layer is pale-brown loamy sand about eight inches thick. The next layer is sandy clay loam about fifty-seven inches thick. . . . Permeability is moderate, and runoff is slow to medium" (*Soil Survey of Florence,* 16, 20).

Appendix D

1. A. Waiters interview, February 20, 1982, p. 92.

2. H. A. Woodle, *Upland Rice Culture in South Carolina* (Clemson, S.C., 1943), 3.

3. Jones, "Upland Rice," 3.

4. Nelson E. Jodon, letter to author, May 21, 1988.

5. Paul Richards, letters to author, April 15, 1988, and September 30, 1992. I am indebted to Richards for his generous sharing of his thoughts and unpublished notes on this subject. Although I give only primary sources for the material presented hereafter, Richards' notes include many of these references.

6. Julian P. Boyd, ed., *The Papers of Thomas Jefferson* (24 vols. completed; Princeton, N.J., 1950–), XIII, 368, XVI, 274, XVII, 564, 5.

7. Richards, letter to author, September 30, 1992; Thomas Jefferson, *Thomas Jefferson's Garden Book, 1766–1824, with relevant extracts from his other writings* (Philadelphia, 1944), 380. I am grateful that Karen Hess sent me material from this book, including a letter that Jefferson wrote to Dr. Benjamin Waterhouse on December 1, 1808. This letter is the basis for the idea that the dryland rice Jefferson received from Africa was an Asian rice, *Oryza mutica.* The letter said: "In answer to the inquiries of the benevolent Dr. De Carro on the subject of the upland or mountain rice, *Oryza Mutica,* I will state to you what I know of it. I first became informed of the existence of a rice which would grow in uplands without anymore water than the common rains, by reading a book of Mr. De Porpre, who had been Governor of the Isle of France, who mentions it as growing there and all along the coast of Africa successfully, and as having been introduced from Cochin-China." I queried T. T. Chang about *Oryza mutica,* and he replied that the library at the International Rice Research Institute could not locate *Oryza mutica* in its computerized files (T. T. Chang, letter to author, October 2, 1990). Richards wrote, "'*Oryza mutica*' is not accepted nomenclature: it was used (presumably) by Jefferson and others in the late 18th century because they thought (wrongly) that 'upland' rice was a distinct species" (Richards, letter to author, September 30, 1992).

8. Boyd, ed., *Papers of Jefferson,* XVIII, 97–98.

9. Jefferson, *Garden Book,* 163. Also see 165.

10. *Ibid.*, 381.
11. *Ibid.;* Richards, letter to author, September 30, 1992.
12. Elaine Nowick, letter to author, May 8, 1989.

APPENDIX E

1. *Encyclopedia Americana,* s.v. *rice.*

2. Charles Gayarré, "Poetry of History of Louisiana," in *Louisiana as It Is,* ed. Dan'[iel] Dennett (New Orleans, 1876), ix–x.

3. Henry P. Dart, "The First Cargo of African Slaves for Louisiana, 1718," trans. Albert Godfrey Sanders, *Louisiana Historical Quarterly,* XIV (April, 1931), 172–73.

4. *Ibid.*, 175–76.

5. Jones, *The Earth Goddess,* 6–7.

6. M. Le Page du Pratz wrote that the Louisiana swamps "bear rice in such plenty, especially the marsh about New Orleans," that the people looked upon the rice as manna ([Antoine Simon] Le Page du Pratz, *The History of Louisiana,* ed. Joseph G. Tregle, Jr. [Baton Rouge, 1975], vii). The use of the word *manna* indicates that the French settlers did view the rice as sent by providence.

7. Lee, "A Culture History of Rice," 89.

8. *Ibid.*, 115; Gayarré, *History of Louisiana,* I, 391. Marcel Giraud wrote that by 1728 there were several mills for hulling rice in the New Orleans area (Marcel Giraud, *The Company of the Indies, 1723–1731,* trans. Brian Pearce [Baton Rouge, 1991], 128, Vol. V of Giraud, *A History of French Louisiana*).

9. Gayarré, "Poetry of History of Louisiana," xii; Amory Austin, *Rice: Its Cultivation, Production, and Distribution in the United States and Foreign Countries* (Washington, D.C., 1893), 9; Littlefield, *Rice and Slaves,* 105; Malcolm Comeaux, "The Environmental Impact," in *The Cajuns: Essays on Their History and Culture,* ed. Glenn R. Conrad (Lafayette, La., 1978), 144.

10. Gabriel Debien, "The Acadians in Santo Domingo, 1764–1789," trans. Glenn R. Conrad, in *The Cajuns,* ed. Conrad, 21, 144.

11. Lauren C. Post, *Cajun Sketches from the Prairies of Southwest Louisiana* (1962; rpr. Baton Rouge, 1990), 33. Post wrote about the peoples in southwest Louisiana—Native Americans, Acadians, African Americans, English Americans, and "another group of some importance result[ing] from the mixing of white people, particularly the French-speaking, with Negroes. The resultant mixing or crossing produced a group of peculiar social position called *Gens Libres de Couleur,* or Free People of Color. Their existence in southwest Louisiana, as well as in other parts of the state, has long been known, but the 'Brand Book for the District of Opelousas and Attakapas, 1760–1888' showed in an impressive way that there were great numbers of these free people of mixed

blood before the Civil War" (33). Consequently, while some Acadians may have learned the providence method of rice cultivation from African American neighbors, others may have learned the method from African American family members.

12. Pete Daniel, *Breaking the Land: The Transformation of Cotton, Tobacco, and Rice Cultures Since 1880* (Urbana, Ill., 1985), 40, 41–42.

13. *Ibid.*, 42. The first expansion of the Louisiana rice industry occurred immediately after the Civil War, when the Louisiana sugar industry lay in ruins and the people turned increasingly to commercial rice production. This commercial production was modeled after the irrigated rice culture that was found along the Louisiana rivers. The second expansion of commercial rice production occurred in the 1880s in the prairie lands of southwestern Louisiana (S. A. Knapp, *The Present Status of Rice Culture in the United States* [Washington, D.C., 1899], 3, 17).

14. At the beginning of the century, James L. Wright, secretary of the American Rice Brokerage Company, wrote: "As a result of the immigration movement started by the Southern Pacific, progressive farmers from the north and west invaded the lands of the quiet Acadians. . . . American ingenuity proceeded to solve the problem of irrigation and the day of 'Providence rice' was past. Providence had made the rice for the simple Acadian folk. . . . But American energy asked only that Providence provide the water, and then went lustily to work to see that they got it" (James L. Wright, "Rice in Texas," *Sunset Magazine* [April, 1906], 585–86). In 1932, Mrs. W. N. Ginn emphasized the importance of the Acadians in her brief history of providence rice by giving thrice as much space to the Acadian people than to the method of cultivation (Mrs. W. N. Ginn, "New History of Rice Industry in America," *Rice Journal*, XXXV [April, 1932], 23).

15. Lee, "A Culture History of Rice," 113.

Appendix F

1. D. Michael Warren and Kristin Cashman, "Indigenous Knowledge for Sustainable Agriculture and Rural Development," *Gatekeeper Series*, No. SA10 (London, 1988), 11–12.

2. A. Waiters interviews, February 20, 1982, pp. 91–95, July 31, 1987, p. 1109.

3. Williamson interview, May 14, 1985, pp. 654–58, 670.

4. H. Waiters interview, August 14, 1987, pp. 1219–26.

5. Note that women dropped the seed and men covered the seed. This may be an Africanism. I questioned Johnson further about this in a telephone conversation. She said that the women would drop the rice seed in a furrow and

her father would plow down the top of the bed to cover the seed. Her brothers never helped drop seed, but they sometimes plowed to cover it. The same procedure was followed with corn and beans (Johnson, telephone conversation with author, June 28, 1988).

6. Johnson interview, March 12, 1984, pp. 185A–185L.

7. Zanders interviews, August 4, 1987, pp. 1198–1202, August 7, 1987, pp. 1206–1208, 1210–11.

Bibliography

Published Works

Abbott, Martin. *The Freedmen's Bureau in South Carolina, 1865–1872.* Chapel Hill, 1967.

Agricultural Extension and Research Liaison Services, Ahmadu Bello University. *Recommended Practices for Swamp Rice Production.* Zaria, Nigeria, 1981.

Andrews, Sidney. *The South Since the War.* New York, 1969.

Archdale, John. "A New Description of That Fertile and Pleasant Province of Carolina." In *Narratives of Early Carolina, 1650–1708,* edited by Alexander S. Salley, Jr. New York, 1911.

Asante, Molefi Kete. "African Elements in African-American English." In *Africanisms in American Culture,* edited by Joseph E. Holloway. Bloomington, Ind., 1990.

Austin, Amory. *Rice: Its Cultivation, Production, and Distribution in the United States and Foreign Countries.* Washington, D.C., 1893.

[Ball, Charles.] *Slavery in the United States: A Narrative of the Life and Adventures of Charles Ball, a Black Man.* 1836; rpr. New York, 1969.

Barber, John W., comp. *A History of the Amistad Captives.* 1840; rpr. New York, 1969.

Bentley, George R. *A History of the Freedmen's Bureau.* New York, 1970.

Berlin, Ira, *et al.*, eds. *The Wartime Genesis of Free Labor: The Lower South.* Cambridge, Eng., 1982. Ser. I, Vol. III of *Freedom: A Documentary History of Emancipation, 1861–1867,* edited by Berlin *et al.* 5 series projected.

Blassingame, John W. *The Slave Community: Plantation Life in the Antebellum South.* New York, 1972.

Bleser, Carol K. Rothrock. *The Promised Land: The History of the South Carolina Land Commission, 1869–1890.* Columbia, S.C., 1969.

Boddie, William Willis. *History of Williamsburg*. Columbia, S.C., 1923.

Bohannan, Paul, and Philip Curtin. *Africa and Africans*. Rev. ed. Garden City, N.Y., 1971.

Boyd, Julian P., ed. *The Papers of Thomas Jefferson*. Vols. XIII, XVI, XVII, and XVIII of 24 vols. completed. Princeton, N.J., 1950–

Brady, Patrick S. "The Slave Trade and Sectionalism in South Carolina, 1787–1808." *Journal of Southern History,* XXXVIII (November, 1972), 601–20.

Cassidy, F. G., and R. B. Le Page, eds. *Dictionary of Jamaican English*. Cambridge, Mass. 1967.

Catesby, Mark. *The Natural History of Carolina, Florida, and the Bahama Islands*. 1731; rpr. Savannah, Ga., 1974.

Chase, Judith Wragg. *Afro-American Art and Craft*. New York, 1971.

Clifton, James M. "The Rice Driver: His Role in Slave Management." *South Carolina Historical Society,* LXXXII (October, 1981), 331–53.

———. "The Rice Industry in Colonial America." *Agricultural History,* LV (July, 1981), 266–83.

———, ed. *Life and Labor on Argyle Island: Letters and Documents of a Savannah River Rice Planter, 1833–1867*. Savannah, Ga., 1978.

Comeaux, Malcolm. "The Environmental Impact." In *The Cajuns: Essays on Their History and Culture,* edited by Glenn R. Conrad. Lafayette, La., 1978.

Cook, Harvey Toliver. *Rambles in the Pee Dee Basin, South Carolina*. Columbia, S.C., 1926.

Cox, LaWanda. "The Promise of Land for the Freedmen." *Mississippi Valley Historical Review,* XLV (December, 1958), 413–40.

Creel, Margaret Washington. *"A Peculiar People": Slave Religion and Community-Culture Among the Gullahs*. New York, 1988.

Current, Richard Nelson. *Those Terrible Carpetbaggers*. New York, 1988.

Curtin, Philip. *The Atlantic Slave Trade: A Census*. Madison, Wis., 1969.

Dalgish, Gerard M. *A Dictionary of Africanisms: Contributions of Sub-Saharan Africa to the English Language*. Westport, Conn., 1982.

Daniel, Pete. *Breaking the Land: The Transformation of Cotton, Tobacco, and Rice Cultures Since 1880*. Urbana, Ill., 1985.

Dart, Henry P. "The First Cargo of African Saves for Louisiana, 1718." Translated by Albert Godfrey Sanders. *Louisiana Historical Quarterly,* XIV (April, 1931), 163–77.

Deas-Moore, Vennie. "Home Remedies, Herb Doctors, and Granny Midwives." *World and I* (January, 1987), 474–85.

Debien, Gabriel. "The Acadians in Santo Domingo, 1764–1789." Translated by Glenn R. Conrad. In *The Cajuns: Essays on Their History and Culture,* edited by Glenn R. Conrad. Lafayette, La., 1978.

De Bow, J. D. B. "Rice." *De Bow's Review,* I (April, 1846), 320–56.

———. "Seacoast Crops of the South." *De Bow's Review,* III (June, 1854), 589–615.

Dennett, John Richard. *The South as It Is, 1865–1866.* Edited by Henry M. Christman. London, 1965.

Dethloff, Henry C. *The History of the American Rice Industry, 1685–1985.* College Station, Tex., 1988.

Doar, David. "Rice and Rice Planting in the South Carolina Low Country." In *Contributions from the Charleston Museum,* edited by E. Milby Burton, VIII (1936), 7–42.

Douglass, Frederick. *Life and Times of Frederick Douglass, Written by Himself: His Early Life as a Slave, His Escape from Bondage, and His Complete History.* Rev. ed. 1892; rpr. New York, 1962.

Drake, St. Clair. *The Redemption of Africa and Black Religion.* Chicago, 1970.

Drayton, John. *A View of South Carolina as Respects Her Natural and Civil Concerns.* Charleston, S.C., 1802.

Du Bois, W. E. Burghardt. *Black Reconstruction in America: An Essay Toward a History of the Part Which Black Folk Played in the Attempt to Reconstruct Democracy in America, 1860–1880.* New York, 1935.

Easterby, J. H., ed. *The South Carolina Rice Plantation as Revealed in the Papers of Robert F. W. Allston.* Chicago, 1945.

Eaton, Clement. *The Growth of Southern Civilization, 1790–1860.* New York, 1961.

Embree, Edwin R., and Julia Waxman. *Investment in People: The Story of the Julius Rosenwald Fund.* New York, 1949.

Evans, DeLancey. "Re-establishment of the Rice Industry in the United States After the Civil War." *Rice Journal,* XXIII (September, 1920), 27–28, 36–37.

Fage, J. D. *A History of West Africa: An Introductory Survey.* 4th ed. Cambridge, Eng., 1969.

Ferguson, Leland. *Uncommon Ground: Archaeology and Early African America, 1650–1800.* Washington, D.C., 1992.

Festinger, Leon. *A Theory of Cognitive Dissonance.* Stanford, 1957.

Foner, Eric. *Nothing but Freedom: Emancipation and Its Legacy.* Baton Rouge, 1983.

———. *Reconstruction: America's Unfinished Revolution, 1863–1877.* New York, 1988.

Franklin, John Hope. *From Slavery to Freedom: A History of American Negroes.* New York, 1947.

Fyle, Clifford N., and Eldred D. Jones. *A Krio-English Dictionary.* Oxford, 1980.

Gayarré, Charles. *History of Louisiana*. Vol. I of 4 vols. 1885; rpr. New York, 1972.

———. "Poetry of History of Louisiana." In *Louisiana as It Is*. Edited by Dan'[iel] Dennett. New Orleans, 1876.

Genovese, Eugene D. *Roll, Jordan, Roll: The World the Slaves Made*. New York, 1974.

Ginn, Mrs. W. N. "New History of Rice Industry in America." *Rice Journal,* XXXV (April, 1932), 23–25.

Giraud, Marcel. *The Company of the Indies, 1723–1731*. Translated by Brian Pearce. Baton Rouge, 1991. Vol. V of Giraud, *A History of French Louisiana*. 5 vols.

Gray, Lewis Cecil. *History of Agriculture in the Southern United States to 1860*. 2 vols. Washington, D.C., 1933.

Gregg, Alexander. *History of the Old Cheraws*. New York, 1867.

Haley, Alex. *Roots: The Saga of an American Family*. New York, 1976.

Hamer, Philip M., and George C. Rogers, Jr., eds., *The Papers of Henry Laurens*. Vols. I and IV of 17 vols. projected. Columbia, S.C., 1968—.

Harllee, William Curry. *Kinfolks: A Genealogical and Biographical Record*. New Orleans, 1934.

Herskovits, Melville J. *Dahomey: An Ancient West African Kingdom*. Vol. II of 2 vols. 1938; rpr. Evanston, Ill., 1967.

———. *The Myth of the Negro Past*. 1941; rpr. Boston, 1972.

———. *The New World Negro: Selected Papers in Afroamerican Studies*. Edited by Frances S. Herskovits. Bloomington, Ind., 1966.

———. "On the Provenience of New World Negroes." *Social Forces,* XII (December, 1933), 247–62.

Herskovits, Melville J., and Frances S. Herskovits. *An Outline of Dahomean Religious Belief*. Menasha, Wis., 1933.

———. *Surinam Folk-Lore*. New York, 1936.

Hess, Karen. *The Carolina Rice Kitchen: The African Connection*. Columbia, S.C., 1992.

Heyward, Duncan Clinch. *Seed from Madagascar*. Chapel Hill, 1937.

Higgins, W. Robert. "The Geographical Origins of Negro Slaves in Colonial South Carolina." *South Atlantic Quarterly,* LXX (Winter, 1971), 34–47.

Holloway, Joseph E. "The Origins of African-American Culture." In *Africanisms in American Culture,* edited by Joseph E. Holloway. Bloomington, Ind., 1990.

Holt, Thomas. *Black over White*. Urbana, Ill., 1977.

Howard, Oliver Otis. *Autobiography of Oliver Otis Howard, Major General United States Army*. New York, 1907.

Hurston, Zora [Neale]. "Hoodoo in America." *Journal of American Folk-Lore,* XLIV (October–December, 1931), 317–417.

Hutchinson, Louise Daniel. *Out of Africa: From West African Kingdom to Colonization*. Washington, D.C., 1979.
Irvine, F. R. *A Text-Book of West African Agriculture: Soils and Crops*. London, 1935.
Jefferson, Thomas. *Thomas Jefferson's Garden Book, 1766–1824, with relevant extracts from his other writings*. Annotated by Edwin Morris Betts. Philadelphia, 1944.
Johnson, Daniel M., and Rex R. Campbell. *Black Migration in America: A Social Demographic History*. Durham, N.C., 1981.
Johnson, Guy B. *Folk Culture on St. Helena Island, South Carolina*. Chapel Hill, 1930.
Johnson, Michael P., and James L. Rourk. *Black Masters: A Free Family of Color in the Old South*. New York, 1984.
Jones, G. Howard. *The Earth Goddess: A Study of Native Farming on the West Coast of Africa*. London, 1936.
Jones, Jenkin W. "Upland Rice." Washington, D.C., 1943. Multigraphed.
Jones, Norrece T., Jr. *Born a Child of Freedom, Yet a Slave: Mechanisms of Control and Strategies of Resistance in Antebellum South Carolina*. Hanover, N.H., 1990.
Joyner, Charles. *Down by the Riverside: A South Carolina Slave Community*. Urbana, Ill., 1984.
King, G. Wayne. *Rise Up So Early: A History of Florence County, South Carolina*. Spartanburg, S.C., 1981.
Knapp, S. A. *The Present Status of Rice Culture in the United States*. Washington, D.C., 1899.
———. "Rice." In *Cyclopedia of American Agriculture*. Vol. II of 4 vols. New York, 1907.
Lawson, Dennis T. *No Heir to Take Its Place: The Story of Rice in Georgetown County, South Carolina*. Georgetown, S.C., 1972.
Le Page du Pratz, [Antoine Simon]. *The History of Louisiana*. Edited by Joseph G. Tregle, Jr. Baton Rouge, 1975.
Levine, Lawrence W. *Black Culture and Black Consciousness: Afro-American Folk Thought from Slavery to Freedom*. New York, 1977.
Levtzion, Nehemia. *Ancient Ghana and Mali*. London, 1973.
Linares de Sapir, Olga. "Agriculture and Diola Society." In *African Food Production Systems,* edited by Peter F. M. McLoughlin. Baltimore, 1970.
Littlefield, Daniel C. *Rice and Slaves: Ethnicity and the Slave Trade in Colonial South Carolina*. Baton Rogue, 1981.
Litwack, Leon F. *Been in the Storm So Long*. New York, 1980.
McCarthy, Louise Miller. *Footprints: The Story of the Greggs of South Carolina*. Winter Park, Fla., 1951.
McCrady, Edward. *The History of South Carolina Under the Proprietary Government, 1670–1719*. New York, 1897.

McDaniel, George W. *Hearth and Home: Preserving a People's Culture.* Philadelphia, 1982.

McPherson, James M. *Ordeal by Fire: The Civil War and Reconstruction.* New York, 1982.

———. *The Struggle for Equality: Abolitionists and the Negro in the Civil War and Reconstruction.* Princeton, 1964.

Magdol, Edward. *A Right to the Land: Essays on the Freedmen's Community.* Westport, Conn., 1977.

Mallard, R. Q. *Plantation Life Before Emancipation.* Richmond, 1892.

Mead, Elwood, ed. *U.S. Department of Agriculture Annual Report of Irrigation and Drainage Investigations, 1904.* Washington, D.C., 1905.

Meriwether, Robert L. *The Expansion of South Carolina, 1729–1765.* Kingsport, Tenn., 1940.

Milling, Chapman J. "Mechanicsville." In *Darlingtoniana: A History of People, Places, and Events in Darlington County, South Carolina,* edited by Eliza Cowan Ervin and Horace Fraser Rudisill. Columbia, S.C., 1964.

Mills, Robert. *Statistics of South Carolina Including a View of Its Natural, Civil, and Military History, General and Particular.* Charleston, S.C., 1826.

M'Tyeire, H. N. *Duties of Christian Masters.* Nashville, 1859.

Mullin, Michael, ed. *American Negro Slavery: A Documentary History.* Columbia, S.C., 1976.

Murdock, George Peter. *Africa: Its Peoples and Their Culture History.* New York, 1959.

Myers, David G. *Social Psychology.* New York, 1983.

Nichols, Elaine, ed. *The Last Miles of the Way: African-American Homegoing Traditions, 1890–Present.* Columbia, S.C., 1989.

Opala, Joseph A. *The Gullah: Rice, Slavery, and the Sierra Leone–American Connection.* Freetown, Sierra Leone, 1987.

Oubre, Claude F. *Forty Acres and a Mule: The Freedmen's Bureau and Black Land Ownership.* Baton Rouge, 1978.

Parrish, Lydia. *Slave Songs of the Georgia Sea Islands.* Hatboro, Pa., 1942.

Pearson, Scott R., J. Dirck Stryker, and Charles P. Humphreys. *Rice in West Africa: Policy and Economics.* Stanford, 1981.

Phillips, Ulrich Bonnell. *American Negro Slavery: Survey of the Supply, Employment, and Control of Negro Labor as Determined by the Plantation Regime.* 1918; rpr. Gloucester, Mass., 1959.

———. *Life and Labor in the Old South.* Boston, 1929.

Poivre, Pierre. *Travels of a Philosopher; or, Observations on the Manners and Arts of Various Nations in Africa and Asia.* Baltimore, 1818.

Post, Lauren C. *Cajun Sketches from the Prairies of Southwest Louisiana.* 1962; rpr. Baton Rouge, 1990.

Ramsay, David. *Ramsay's History of South Carolina, from Its First Settlement in 1670 to the Year 1808.* Vol. II of 2 vols. 1809; rpr. Newberry, S.C., 1858.

Rawick, George P., ed. *The American Slave: A Composite Autobiography.* Vol. II of 19 vols. 1941; rpr. Westport, Conn., 1972.

Richards, Paul. *Coping with Hunger: Hazard and Experiment in an African Rice-Farming System.* London, 1986.

Rodney, Walter. *A History of the Upper Guinea Coast, 1545–1800.* London, 1970.

Rogers, George C., Jr. *The History of Georgetown County, South Carolina.* Columbia, S.C., 1970.

Rose, Willie Lee. "Jubilee & Beyond: What Was Freedom?" In *What Was Freedom's Price?,* edited by David G. Sansing. Jackson, Miss., 1978.

———. *Rehearsal for Reconstruction: The Port Royal Experiment.* Indianapolis, 1964.

———. *Slavery and Freedom.* Edited by William W. Freehling. New York, 1982.

Rosengarten, Dale. "Lowcountry Baskets: A Place in the Sun." *Carologue* (January–February, 1987), 1, 8–9.

———. *Row upon Row: Seagrass Baskets of the South Carolina Lowcountry.* Columbia, S.C., 1986.

Rosengarten, Theodore. *The Life of Nate Shaw.* New York, 1974.

———. *Tombee: Portrait of a Cotton Planter, with the Journal of Thomas B. Chaplin (1822–1890).* New York, 1986.

Rothenberg, Paula S. *Racism and Sexism: An Integrated Study.* New York, 1988.

Ruthenberg, Hans. *Farming Systems in the Tropics.* 2nd ed. Oxford, 1976.

Salley, A. S. *The Introduction of Rice Culture into South Carolina.* In Bulletin of the Historical Commission of South Carolina, No. 6. Columbia, S.C., 1919.

———. "The True Story of How Madagascar Gold Seed Was Introduced into South Carolina." In *Contributions from the Charleston Museum,* edited by E. Milby Burton, VIII (1936), 51–53.

Sass, Herbert Ravenel. "The Rice Coast: Its Story and Its Meaning." In Alice Ravenel Huger Smith, *A Carolina Rice Plantation in the Fifties: Thirty Paintings in Water-Colour.* New York, 1936.

Sellers, W. W. *A History of Marion County, South Carolina.* Columbia, S.C., 1902.

Simkins, Francis Butler, and Robert Hilliard Woody. *South Carolina During Reconstruction.* Chapel Hill, 1932.

Smith, Alice Ravenel Huger. *A Carolina Rice Plantation in the Fifties: Thirty Paintings in Water-Colour.* New York, 1936.

Smith, Julia Floyd. *Slavery and Rice Culture in Low Country Georgia, 1750–1860.* Knoxville, 1985.

Sobel, Mechal. *Trabelin On: The Slave Journey to American Afro-Baptist Faith.* Westport, Conn., 1979.

Soil Survey of Florence and Sumter Counties, South Carolina. Washington, D.C., 1974.

South Carolina. State Board of Agriculture. *South Carolina Resources and Population, Institutions and Industries.* Charleston, S.C., 1883.

Stampp, Kenneth M. *The Era of Reconstruction, 1865–1877.* New York, 1966.

———. *The Peculiar Institution.* New York, 1956.

Stroyer, Jacob. *My Life in the South.* 3rd ed. Salem, 1885.

Stuckey, Sterling. *Slave Culture: Nationalist Theory and the Foundations of Black America.* New York, 1987.

Taylor, Alrutheus Ambush. *The Negro in South Carolina During Reconstruction.* New York, 1924.

Taylor, Joe Gray. *Negro Slavery in Louisiana.* Baton Rouge, 1963.

Thompson, Robert Farris. *Flash of the Spirit: African and Afro-American Art and Philosophy.* New York, 1983.

Trefousse, Hans L. *Carl Schurz: A Biography.* Knoxville, 1982.

Turner, Lorenzo D. *Africanisms in the Gullah Dialect.* 1949; rpr. New York, 1969.

Vass, Winifred Kellersberger. *The Bantu Speaking Heritage of the United States.* Los Angeles, 1979.

Wallace, David Duncan. *South Carolina: A Short History, 1520–1948.* Columbia, S.C., 1951.

Warren, D. Michael, and Kristin Cashman. "Indigenous Knowledge for Sustainable Agriculture and Rural Development." *Gatekeeper Series,* No. SA10. London, 1988.

Werner, M. R. *Julius Rosenwald: The Life of a Practical Humanitarian.* New York, 1939.

West, Robert C. *The Pacific Lowlands of Colombia: A Negroid Area of the American Tropics.* Baton Rouge, 1957.

Williamson, Joel. *After Slavery: The Negro in South Carolina During Reconstruction, 1861–1877.* Chapel Hill, 1965.

Wood, Peter. *Black Majority: Negroes in Colonial South Carolina from 1670 Through the Stono Rebellion.* 1974; rpr. New York, 1975.

Wood, Peter H., and Karen C. C. Dalton. *Winslow Homer's Images of Blacks: The Civil War and Reconstruction Years.* Austin, 1988.

Woodle, H. A. *Upland Rice Culture in South Carolina.* Clemson, S.C., 1943.

Woodson, Carter G. *The Negro in Our History.* 5th ed. Washington, D.C., 1928.

Wright, James L. "Rice in Texas." *Sunset Magazine.* April, 1906, pp. 584–89.

Wrigley, Gordon. *Tropical Agriculture: The Development of Production.* London, 1971.

Interviews with Author

Tapes and transcripts of the following interviews will be deposited in the Caroliniana Library, University of South Carolina, Columbia, S.C.

Bailey, Willie H., July 7, 1988.
Coker, Annie, May 13, 20, 1985.
Coker, Leon H., March 14, 1984.
Egleton, Maggie Waiters, March 3, 1986.
Gregg, Mattie Smalls, August 3, 1987.
Johnson, Frances, March 12, July 2, 1984.
Peterson, Rubin, February 8, 1986.
Robinson, Annie Lee Waiters, March 23, 1984.
Saunders, Frank, August 17, 1987.
Sellers, Mabel Smalls, February 27, 1984.
Taylor, Harmon, July 31, 1987.
Waiters, Archie, November, 1975, October 15, 1976, December 28, 29, 31, 1977, September 17, 1980, February 5, 20, July 20, 1982, March 23, June 5, 26, 28, July 19, 23, 25, 1984, June 5, 6, 12, 1985, February 4, 26, March 1, 1986, July 31, October 15, 1987, June 28, July 31, 1988, March 9, 1990.
Waiters, Archie, and Catherine Waiters (joint interviews), December 29, 1977, February 4, 1986, March 9, 1990.
Waiters, Catherine, October 15, 1976, December 29, 31, 1977, March 23, 1984, June 12, 1985, February 4, 19, 1986, July 31, October 15, 1987, June 6, 29, 1988, December 29, 1989, March 9, 1990.
Waiters, Hester, August 14, 1987.
Washington, Mary Daniels, February 23, 1984.
Williams, Claudia Williamson, July 17, 1984.
Williams, Dorothy Smalls, August 16, 1987.
Williamson, Matthew, May 14, 1985.
Zanders, Ida Ellison, August 4, 7, 1987, June 30, 1988.

Personal Communications

Notes from the conversations listed below are filed alphabetically in conversation volume of interview notebooks, to be deposited in the Caroliniana Library, University of South Carolina, Columbia, S.C.

Bah, Alpha. Conversation with author, Charleston, S.C., September 18, 1990.

Bailey, Ruth Lee Scott. Conversation with author, Mars Bluff, S.C., July 7, 1988.

Broadwell, Joseph. Conversation with author, Mars Bluff, S.C., May 12, 1984.

Carter, Hazel. Letter to author, December 2, 1990.

Chang, T. T. Letter to author, October 2, 1990.

Dargan, Timothy G. Conversation with author, Mars Bluff, S.C., July 11, 1988.

Harwell, James R. Letter to author, June, 1984.

Harwell, Lacy Rankin. Conversation with author, Mars Bluff, S.C., July 16, 1984.

Jodon, Nelson E. Letter to author, May 21, 1988.

Leach, Melissa. Letter to author, May 23, 1989.

Nowick, Elaine. Letter to author, May 8, 1989.

Opala, Joseph A. Conversation with author, Orangeburg, S.C., September 7, 1990.

Pinkney, Atleene. Letter to author, March, 1988.

Richards, Paul. Conversation with author, Ames, Iowa, April 14, 1989.

———. Letters to author, April 15, 1988, [March?], 1989, September 30, 1992.

Rudisill, Horace Fraser. Letter to author, May 18, 1988.

Schulze, Richard R. Letter to author, January 9, 1989.

Sengova, Joko. Conversation with author, Charleston, S.C., September 18, 1990.

Smith, Isabell Daniels. Telephone conversation with author, March 13, 1984.

Swails, Lawrence. Conversation with author, Mars Bluff, S.C., June 29, 1992.

Vernon, Jane. Letters to author, August 24, 1988, September 19, 1988.

Waiters, Catherine. Letters to author, November, 11, 1985, May 28, 1990.

Waiters, Otis. Letter to author, October 29, 1991.

Williamson, Ann Niemeyer. Letter to author, May 22, 1992.

Williamson, Juanita Cody. Letter to author, November 17, 1988.

Zanders, Ida. Letter to author, December 6, 1987.

Unpublished Materials

Bennett, John. Undated file of notes (*ca.* 1900–30), File 21-111, Folder 20. South Carolina State Historical Society, Charleston, S.C.

Dargan, Timothy G. "Canals Leading to Irrigated Rice Fields at Mars Bluff." Map drawn for author, July 11, 1988, in author's conversation notebook, to be deposited at the Caroliniana Library, University of South Carolina, Columbia, S.C.

Fraser, John. Papers. In possession of Horace Fraser Rudisill, Darlington, S.C.

Gregg, J. Eli. Gregg and Son, Mars Bluff, Cotton and Corn Records, 1860–1867. Bound MS vol. South Caroliniana Library, Columbia, S.C.

Gregg, Walter. Letters to Anna Parker Gregg, October 11, 28, December 26, 1867. Photocopied and compiled by Anna S. Sherman, January, 1985.

Law, William. Papers, 1761–1890, Darlington County. Manuscript Department, William R. Perkins Library, Duke University, Durham, N.C.

Miller, John Blount. Papers. Box 2, Folder 2. Manuscript Department, William R. Perkins Library, Duke University, Durham, N.C.

"The Old Slave Mart Museum." One-page flyer. N.d. South Carolina State Historical Society, Charleston, S.C.

Pinckney, Henry L. Plantation Book, 1850–1867. Manuscript Department, William R. Perkins Library, Duke University, Durham, N.C.

Pugh, Evan. "The Private Journal of Evan Pugh," as condensed and typed by Mr. and Mrs. O. L. Warr. In possession of Darlington County Historical Commission, Darlington, S.C.

Williamson, George Lawrence. Journal. In possession of Juanita Cody Williamson, Florence, S.C.

Public Documents

Bostick, Eli M. Easement to William R. Johnson, August 3, 1846. Marion County, S.C., Deed Book T. South Carolina Department of Archives and History, Columbia, S.C.

Bostick, Tristram. Appraisement of Estate, January 14, 1820. Marion County Court House, Marion, S.C.

Brown, John A. Easement to W. R. Johnson and W. T. Wilson, 1850. Marion County, S.C., Deed Book W. South Carolina Department of Archives and History, Columbia, S.C.

Commissioners of Sinking Fund. Deed to Washington James *et al.*, March 4, 1884. Marion County, S.C., Deed Book MM. Records of the Budget and Control Board; Sinking Fund Commission, Public Land Division, Duplicate Titles, B. South Carolina Department of Archives and History, Columbia, S.C.

Gregg, James. Appraisement of Estate, March 16, 1802. Marion County Court House, Marion, S.C.

Gregg, John. Appraisement of Estate, October 24, 1839. Marion County Court House, Marion, S.C.

Grice, Sarah, and J. A. Grice. Deed to Sidney James *et al.*, March 26, 1891. Florence County, S.C., Deed Book E. South Carolina Department of Archives and History, Columbia, S.C.

Howard, Anthony H. Deeds, 1867, 1871, and 1875. Marion County, S.C.,

Deed Books EE, FF, and GG. Marion County Courthouse, Marion, S.C.
Johnson, W. R. Plat, 1844. Marion County, S.C., Plat Book B. South Carolina Department of Archives and History, Columbia, S.C.
Lane, James. Appraisement of Estate, October 7, 1844. Marion County Court House, Marion, S.C.
———. Deed to Dr. William R. Johnson, May 20, 1844. Marion County, S.C., Deed Book S. South Carolina Department of Archives and History, Columbia, S.C.
McIsick, Eli, and Mary E. Poston. Deed to Ervin James, January 23, 1871. Marion County, S.C., Deed Book DD. South Carolina Department of Archives and History, Columbia, S.C.
Napier, Robert. Plat, November 3, 1846. Marion County, S.C., Plat Book B. South Carolina Department of Archives and History, Columbia, S.C.
Pearce, Robert H., *et al.* Deed to Dr. William R. Johnson, December 28, 1859. Marion County, S.C., Deed Book Y. South Carolina Department of Archives and History, Columbia, S.C.
Thompson, Hannah. Will, October 21, 1856. Marion County Court House, Marion, S.C.
United States Census, 1850. Schedule 2, Slave Inhabitants in the County of Marion, State of South Carolina. South Carolina Department of Archives and History, Columbia, S.C.
U.S. Census: Original Agriculture, Industry, Social Statistics, and Mortality Schedules for South Carolina, 1850. Schedule 4, Productions of Agriculture in Marion District, South Carolina. South Carolina Archives Microcopy No. 2. South Carolina Department of Archives and History, Columbia, S.C.
Wilson, William T. Easement to William R. Johnson, August 3, 1846. Marion County, S.C., Deed Book T. South Carolina Department of Archives and History, Columbia, S.C.

Typescripts, Theses, and Dissertations

Carney, Judith Ann. "The Social History of Gambian Rice Production: An Analysis of Food Security Strategies." Ph.D. dissertation, University of California, Berkeley, 1986.
Coon, David LeRoy. "The Development of Market Agriculture in South Carolina, 1670–1785." Ph.D. dissertation, University of Illinois, 1972.
Davis, Henry E. "A History of Florence, City and County, and of Portions of the Pee Dee Valley, South Carolina." Typescript, 1965. Special Collections, Francis Marion College Library, Florence, S.C.
Lee, Chan. "A Culture History of Rice with Special Reference to Louisiana." Ph.D. dissertation, Louisiana State University, 1960.

Seagrave, Charles Edwin. "The Southern Negro Agricultural Worker, 1850–1870." Ph.D. dissertation, Stanford University, 1971.

OTHER

West, Alex. "The Strength of These Arms: Black Labor—White Rice." Video. Durham, N.C., 1988.

Index

Page numbers in italics refer to photographs and maps.